# UNDERSTANDING
# CONSTRUCTION
# DRAWINGS
## Fifth Edition

unit 2 answers
questions 1-10
FRIDAY

TEST through
chapter 1-8
FRIDAY NEXT WEEK

Don Mulrooney
08/22/13

# UNDERSTANDING CONSTRUCTION DRAWINGS
## Fifth Edition

**MARK W. HUTH**

**DELMAR**
**CENGAGE Learning**

Australia • Brazil • Japan • Korea • Mexico • Singapore • Spain • United Kingdom • United States

**DELMAR**
CENGAGE Learning™

**Understanding Construction Drawings,
Fifth Edition**
Mark W. Huth

Vice President, Editorial: Dave Garza

Director of Learning Solutions: Sandy Clark

Senior Acquisitions Editor: James Devoe

Managing Editor: Larry Main

Senior Product Manager: Sharon Chambliss

Editorial Assistant: Cristopher Savino

Vice President, Marketing: Jennifer McAvey

Executive Marketing Manger:
  Deborah S. Yarnell

Marketing Manager: Jimmy Stephens

Marketing Specialist: Mark Pierro

Production Director: Wendy Troeger

Production Manager: Mark Bernard

Content Project Manager: Michael Tubbert

Art Director: Bethany Casey

For product information and technology assistance, contact us at
**Professional & Career Group Customer Support, 1-800-648-7450**

For permission to use material from this text or product,
submit all requests online at **cengage.com/permissions**
Further permissions questions can be e-mailed to
**permissionrequest@cengage.com**

Library of Congress Control Number: 2008942495
ISBN-13: 978-1-4354-6447-6
ISBN-10: 1-4354-6447-8

**Delmar**
5 Maxwell Drive
Clifton Park, NY 12065-2919
USA

Cengage Learning products are represented in Canada by Nelson Education, Ltd.

For your lifelong learning solutions, visit **delmar.cengage.com**
Visit our corporate website at **cengage.com**

**Notice to the Reader**

Printed in the United States of America
2 3 4 5 XX 11 10

# Contents

# Preface

## Intended Audience

*Understanding Construction Drawings* is intended for students in construction programs in two-year and four-year colleges and technical institutes, as well as apprentice training. Designed for a course in print reading focused on both residential and commercial construction, the book helps you learn to read the drawings that are used to communicate information about buildings. It includes drawings for buildings that were designed for construction in several parts of North America. The diversity of building classifications and geographic locations ensures that you are ready to work on construction jobs anywhere in the industry. Everyone who works in building construction should be able to read and understand the drawings of the major trades.

## How to Use This Book

The book is divided into four major parts and several units within each part. Each part relates to the prints in the separate drawing packet,

**Part I, Drawings—The Language of Construction,** introduces you to the basics of print reading by covering views, scales, lines, and symbols, as well as the various plan views, elevations, and sections and details.

**Part II, Reading Drawings for Trade Information,** provides information on how to interpret drawings for project specifics—everything from footings and foundation walls to room finishing and cabinets.

**Part III, Multifamily Construction,** dives into the details of more advanced residential print reading and applies the skills learned in Parts I and II to other types of construction, as well as mechanical and electrical trades.

**Part IV, Heavy Commercial Construction,** presents the need-to-know information on interpreting prints for large commercial construction including structural drawings, mechanical drawings, and electrical drawings.

## Features of the Units

The individual units are made up of four elements: Objectives, the main body of the unit, Check Your Progress, and Assignment.

○ The **Objectives** appear at the beginning of the unit so that you will know what to look for as you study the unit.

○ The **main body** is the presentation of content with many illustrations and references to the prints for the building being studied in that part.

○ **Check Your Progress** gives you a quick check on how well you understood the main points in the unit.

Each unit contains ten to twenty *Assignment* questions that require you both to understand the content of the unit and to apply that understanding to reading the drawings. There are more than 600 questions in the book.

## The Drawing Packet, Glossary, and Appendix

At the back of the text you will find several helpful aids for studying construction drawings.

○ A **drawing packet** conveniently boxed with the book contains four separate plans that relate to each of the parts within the book. The drawing packet contains prints for four buildings: a simple two-family duplex that is very easy to understand, a more complex single-family home, one unit of a townhouse that uses different materials and methods than the first two buildings, and an addition to a school.

O The *Glossary* defines all the new technical terms introduced throughout the textbook. Each of these terms is defined where it is first used, but if you need to refresh your memory, turn to the Glossary.

O The *Math Reviews* in the Appendix are an innovative feature that has helped many construction students through a difficult area. These are concise reviews of the basic math you are likely to encounter throughout the building construction field. As math is required in this textbook, reference is made to the appropriate Math Review. All the math skills needed to complete the end-of-unit assignments in this book are covered in these Math Reviews.

O The Appendix also includes a complete list of construction abbreviations used on the prints, along with their meaning.

O Another section of the Appendix explains the most commonly used symbols for materials and small equipment.

## New to This Edition

*Understanding Construction Drawings,* Fifth Edition, includes up-to-date information on the latest in the industry, as well as a new look. The book has been updated to the *International Building Code, 2006 Edition* and contains references to this standard for the building trades

## Supplements to the Text

An *Instructor's Guide* gives answers to all the Assignment questions in the textbook and explains how the answers were found or calculated. The *Instructor's Guide* also contains more than 500 additional questions that can be used for test, supplemental assignments, and review. The answer to each of these questions is given, along with an explanation of the answer. *(Order number: 1435464486)*

## Acknowledgments

I am grateful to all who contributed to this textbook. The students who studied from and the instructors who taught from the first four editions and offered a wealth of valuable feedback played an instrumental role in shaping this edition. Several companies provided expertise and contributed illustrations. They are named in the captions to the illustrations they provided. I would especially like to thank the architects and engineers who supplied the drawings for the drawing packet:

Robert Kurzon, A.I.A., for the Duplex and Lake House

Carl Griffith, A.I.A., at Cataldo, Waters, and Griffith Architects, P.C., and HA2F Consultants in Engineering for the School Addition

I would also like to thank the instructors who reviewed the manuscript for the previous editions and for this new edition, providing guidance in making it the best print reading textbook it could be.

Noah Bartnick
North Dakota State College of Science
Wahpeton, ND

Jay Boyle, Registered Architect
Architectural Technology Instructor
Anoka-Hennepin Technical College
Anoka, MN

Joseph Coleman
University of Cincinnati
Cincinnati, OH

Judson W. Collins
Self-employed
Edmund, OK

Joseph Dusek
Professor of Construction Management
Triton College
River Grove, IL

William Earlton
Los Angeles Trade Technical College
Los Angeles, CA

Earl Garrick
Pima Community College
Tucson, AZ

Charles Gordon
Retired plumbing contractor
Jacksonville, FL

Michael Joyce
Joyce Plumbing Inc.
Cary, NC

Larry Kness
Southeast Community College
Milford, NE

Paul Litchy
Nashville Tech College
Nashville, TN

Greg Pakieser
Joliet Junior College
Joliet, IL

Cathy Rock
Red Rocks Community College
Denver, CO

Wayne Sheppard
Pennsylvania College of Technology
Williamsport, PA

John Steele, P.E.
Steele Engineering and Consulting, P.C.
Troy, NY

Marty Swan
Lincoln Land Community College
Springfield, IL

## About the Author

The author of this textbook, Mark W. Huth, brings many years of experience in the industry to his writing—first working as a carpenter, then a contractor, and then a building construction teacher—and his career has allowed him to consult with hundreds of construction educators in high schools, colleges, and universities. He has also authored several other successful construction titles, including *Basic Blueprint Reading for Construction, Residential Construction Academy: Basic Principles for Construction,* and *Practical Problems in Mathematics for Carpenters.*

# DRAWINGS—THE LANGUAGE OF CONSTRUCTION

Part I helps you develop a foundation upon which to build skills and knowledge in reading the drawings used in the construction industry. The topics of the various units in this section are the basic concepts upon which all construction drawings are read and interpreted. The details of construction are explored in Parts II, III, and IV.

Many of the assignment questions in this part refer to the drawings of the Duplex included in the drawing packet that accompanies this textbook. The Duplex was designed as income property for a small investor. It was built on a corner lot in a small city in upstate New York. The Duplex is an easy-to-understand building. Its one-story, rectangular design requires only a minimum of views; you can quickly become familiar with the Duplex drawings.

# The Design-Construction Sequence and the Design Professions

## Objectives

After completing this unit, you will be able to perform the following tasks:

O Name the professions included in the design and planning of a house or light commercial building.

O List the major functions of each of these professions in the design and planning process.

O Identify the profession or agency that should be contacted for specific information about a building under construction.

The construction industry employs about 15 percent of the working people in the United States and Canada. More than 60 percent of these workers are involved in new construction. The rest are involved in repairing, remodeling, and maintenance. As the needs of our society change, the demand for different kinds of construction increases. Homeowners and businesses demand more energy-efficient buildings. The shift toward automation in business and industry means that new offices are needed. Our national centers of commerce and industry are shifting. These are only a few of the reasons that new housing starts are considered important indicators of our economic health.

There are four main classifications of construction: residential, commercial, industrial, and civil. *Residential construction* includes single-family homes, small apartment buildings, and condominiums, **Figure 1–1(A)**. *Commercial construction* includes office and apartment buildings, hotels, stores, shopping centers, and other large buildings, **Figure 1–1(B)**. *Industrial construction* includes structures other than buildings, such as refineries and paper mills, that are built for industry, **Figure 1–1(C)**. *Civil construction,* **Figure 1–1(D),**

**Figure 1–1(A).** Residential construction.

**Figure 1–1(B).** Commercial construction.

**Figure 1–1(C).** Industrial construction. Delta Energy Center, water treatment tanks and buildings in the foreground. *Courtesy of Bechtel Corporation. Photographer: Terry Lowenthal. Used by permission.*

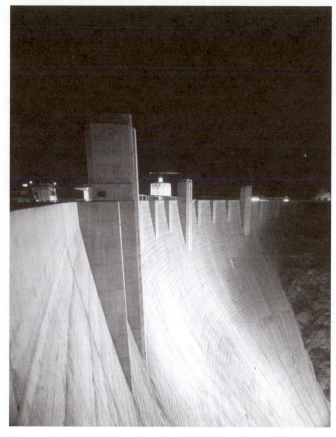

**Figure 1–1(D).** Civil construction. At 726 feet, Hoover Dam is the highest dam in the United States. *Courtesy of Bechtel Corporation. Photographer: Ray Frayne. Used by permission.*

is more closely linked with the land and refers to highways, bridges, airports, and dams, for example.

## The Design Process

The design process starts with the owner. The owner has definite ideas about what is needed but may not be expert at describing that need or desire in terms the builder can understand. The owner contacts an architect to help plan the building.

The architect serves as the owner's agent throughout the design and construction process. Architects combine their knowledge of construction—of both the mechanics and the business—with artistic or aesthetic knowledge and ability. They design buildings for appearance and use.

The architect helps the owner determine how much space is needed, how many rooms are needed for now and in the future, what type of building best suits the owner's lifestyle or business needs, and what the costs will be. As the owner's needs take shape, the architect makes rough sketches to describe the planned building. At first these may be balloon diagrams, **Figure 1–2,** to show traffic flow and the number of rooms. Eventually, the design of the building begins to take shape, **Figure 1–3.**

Before all the details of the design can be finalized, other construction professionals become involved. Building codes specify requirements to ensure that buildings are safe from fire hazards, earthquakes, termites, surface water, and other concerns of the community. Most building codes are based on a model code. For example, the International Code Council (I.C.C.) publishes several model codes, one of which is the *International Residential Code® for One- and*

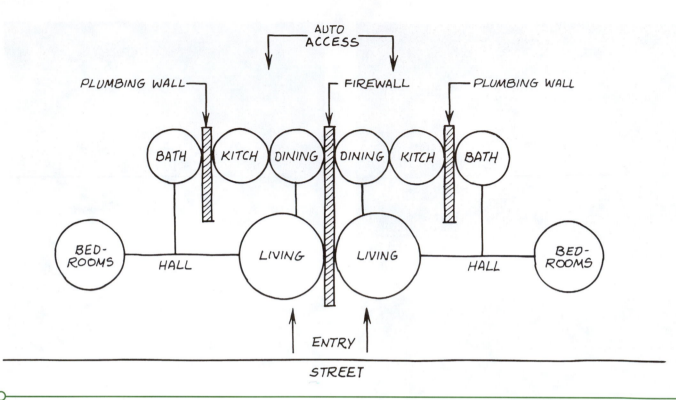

**Figure 1–2.** Balloon sketch of Duplex.

**Figure 1–3.** Straight line sketch of Duplex.

*Two-Family Dwellings,* **Figure 1–4.** This often is called simply the Residential Code. It is a model code, because it is a model that may be used by state and local building authorities as a basis for their own local code. A model code has no authority on its own. The government having jurisdiction in a locale must adopt its own building code. Very often the government body having jurisdiction (called the *Authority Having Jurisdiction,* or AHJ) adopts the model code. Sometimes the AHJ adds specific clauses to the model, and, in rare cases, it writes an all-new code. State building codes allow local governments to adopt a local building code, but they require that the local code be at least as stringent as the state code.

**Figure 1–5.** Stock plans can be ordered from catalogs.

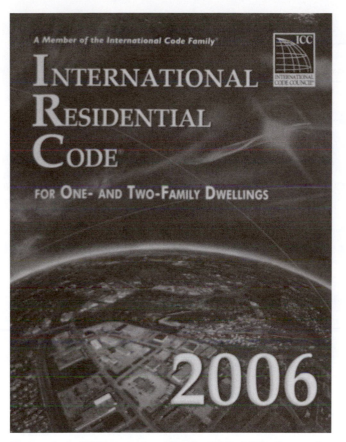

**Figure 1–4.** *2006 International Residential Code® for One- and Two-Family Dwellings.*

The local building code is administered by a building department of the local government. The building department reviews the architect's plans before construction begins and inspects the construction throughout its progress to ensure that the code is followed.

Most communities also have zoning laws. *A zoning law* divides the community into zones where only certain types of buildings are permitted. Zoning laws prevent such problems as factories and shopping centers being built in the same neighborhood as homes.

Building departments usually require that very specific procedures are followed for each construction project. A building permit is required before construction begins. The building permit notifies the building department about planned construction. Then, the building department can make sure that the building complies with all the local zoning laws and building codes. When the building department approves the completed construction,

it issues a *certificate of occupancy*. This certificate is not issued until the building department is satisfied that the construction has been completed according to the local code. The owner is not permitted to move into the new building until the certificate of occupancy has been issued.

If the building is more complex than a home or simple frame building, engineers may be hired to help design the structural, mechanical, electrical, or other aspects of the building. Consulting engineers specialize in certain aspects of construction and are employed by architects to provide specific services. Finally, architects and their consultants prepare construction drawings that show all aspects of the building. These drawings tell the contractor specifically what to build.

Many homes are built from stock plans available from catalogs of house designs, building materials dealers, or magazines, **Figure 1–5.** However, many states require a registered architect to approve the design and supervise the construction.

## Starting Construction

After the architect and the owner decide on a final design, the owner obtains financing. The most common way of financing a home is through a mortgage. A *mortgage* is a guarantee that the loan will be paid in installments. If the loan is not paid, the lender has the right to sell the building in order to recover the money owed. In return for the use of the lender's money, the borrower pays interest—a percentage of the outstanding balance of the loan.

When financing has been arranged (sometimes before it is finalized), a contractor is hired. Usually a

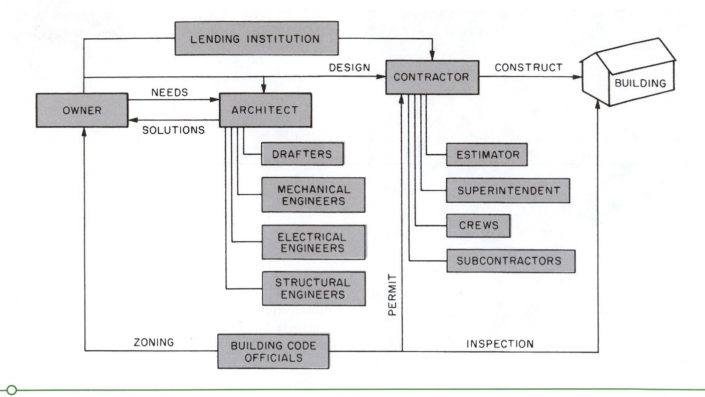

**Figure 1–6.** Design and construction team.

general contractor is hired with overall responsibility for completing the project. The general contractor in turn hires subcontractors to complete certain parts of the project. All stages of construction may be subcontracted. The parts of home construction most often subcontracted are plumbing and heating, electrical, drywall, painting and decorating, and landscaping. The relationships of all the members of the design and construction team are shown in **Figure 1–6.** Utility installers should carefully investigate all the drawings, especially the architectural drawings, in order to determine the installation locations of their equipment.

## ✔ CHECK YOUR PROGRESS

Can you perform these tasks?

● List construction design professions.

● Describe what work is done by each of these professions.

● Name the profession responsible for each major part of the design-construction progress.

## ASSIGNMENT

1. Who acts as the owner's agent while the building is being constructed?

2. Who designs the structural aspects of a commercial building?

3. Who would normally hire an electrical engineer for the design of a store?

4. Who is generally responsible for obtaining financing for a small building?

5. To whom would the general contractor go if there were a problem with the foundation design for a home?

6. If local building codes require specific features for earthquake protection, who is responsible for seeing that they are included in a home design?

7. Whom would the owner inform about last-minute changes in the interior trim when the building is under construction?

8. What regulations specify what parts of the community are to be reserved for single-family homes only?

9. Who issues the building permit?

10. What regulations are intended to ensure that all new construction is safe?

# UNIT 2

# Views

## Objectives

After completing this unit, you will be able to perform the following tasks:

○ Recognize oblique, isometric, and orthographic drawings.

○ Draw simple isometric sketches.

○ Identify plan views, elevations, and sections.

## Isometric Drawings

A useful type of pictorial drawing for construction purposes is the **isometric drawing.** In an isometric drawing, vertical lines are drawn vertically, and horizontal lines are drawn at an angle of 30° from horizontal, **Figure 2–1.** All lines on one of these isometric axes are drawn in proportion to their actual length. Isometric drawings tend to look out of proportion because we are used to seeing the object appear smaller as it gets farther away.

Isometric drawings are often used to show plumbing, **Figure 2–2.** The ability to draw simple isometric sketches is a useful skill for communicating on the job site. Try sketching a brick in isometric as shown in **Figure 2–3.**

Step 1.    Sketch a Y with the top lines about 30° from horizontal.

Step 2.    Sketch the bottom edges parallel to the top edges.

Step 3.    Mark off the width on the left top and bottom edges. This will be about twice the height.

Step 4.    Mark off the length on the right top and bottom edges. The length will be about twice the width.

Step 5.    Sketch the two remaining vertical lines and the back edges.

**Figure 2–1.** Isometric of building.

Figure 2–2. Single-line plumbing isometric.

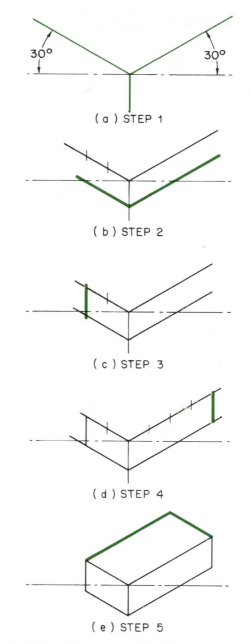

Figure 2–3. Sketching an isometric brick.

Other isometric shapes can be sketched by adding to or subtracting from this basic isometric brick, **Figure 2–4.** Angled surfaces are sketched by locating their edges and then connecting them.

## Oblique Drawings

When an irregular shape is to be shown in a pictorial drawing, an **oblique drawing** may be best. In oblique drawings, the most irregular surface is drawn in proportion as though it were flat against the drawing surface. Parallel lines are added to show the depth of the drawing, **Figure 2–5.**

## Orthographic Projection

To show all information accurately and to keep all lines and angles in proportion, most construction drawings are drawn by **orthographic projection**. Orthographic projection is most often explained by imagining the object to be drawn inside a glass box. The corners and the lines representing the edges of the object are then projected onto the sides of the box, **Figure 2–6.** If the box is unfolded, the images projected onto its sides will be on a single plane, as on a sheet of paper, **Figure 2–7.** In other

(a) GABLE ROOF BUILDING

(b) ELL-SHAPED BUILDING

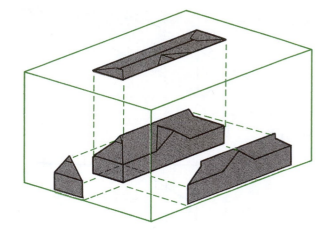

**Figure 2–6.** Duplex inside a glass box; method of orthographic projection of roof, front side, and end.

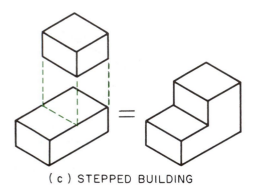

(c) STEPPED BUILDING

**Figure 2–4.** Variations on the isometric brick.

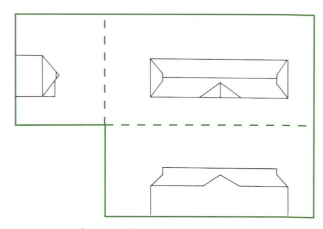

**Figure 2–7.** Orthographic projection unfolded on a flat sheet of paper.

(a)    FRONT VIEW OF CROWN MOLDING

(b)    OBLIQUE VIEW OF CROWN MOLDING

**Figure 2–5.** Oblique drawing.

words, in orthographic projection each view of an object shows only one side (or top or bottom) of the object.

All surfaces that are parallel to the plane of projection (the surface of the box) are shown in proportion to their actual size and shape. However, surfaces that are not parallel to the plane of projection are not shown in proportion. For example, both of the roofs in the top views of **Figure 2–8** appear to be the same size and shape, but they are quite different. To find the actual shape of the roof, you must look at the end view.

In construction drawings, the views are called plans and elevations. A *plan view* shows the layout of the object as viewed from above, **Figure 2–9**. A set of drawings for a building usually includes plan views of the site (lot), the floor layout, and the foundation. **Elevations** are drawings that show height. For example, a drawing that

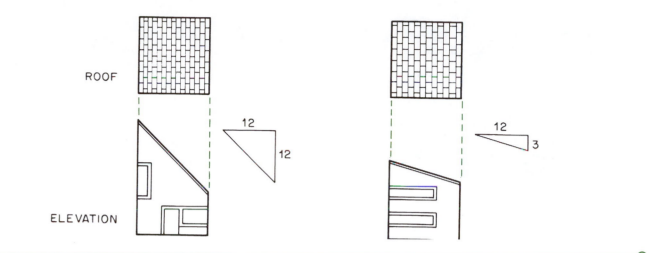

**Figure 2–8.** Views of two shed roofs.

(a) VIEW OF SITE

(b) PLOT PLAN

**Figure 2–9.** Plan view.

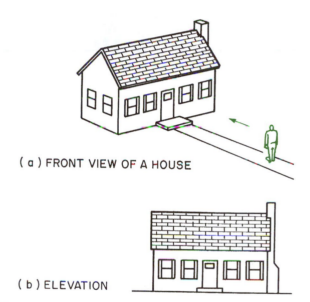

(a) FRONT VIEW OF A HOUSE

(b) ELEVATION

**Figure 2–10.** Building elevation.

shows what would be seen standing in front of a house is a building elevation, **Figure 2–10.** Elevations are also used to show cabinets and interior features.

Because not all features of construction can be seen in plan views and elevations from the outside of a building, many construction drawings are **section views.** A section view, usually referred to simply as a *section,* shows what would be exposed if a cut were made through the object, **Figure 2–11.** Actually, a floor plan is a type of section view, **Figure 2–12.** It is called a *plan* because it is in that position – viewed from above – but it is a type of *section* because it shows what would be exposed if a cut were made through the building. Most section views are called sections, but floor plans are customarily referred to as plans or floor plans.

(a)

( b ) SECTION

**Figure 2–11.** Section of a window sash.

(a)

(b)

**Figure 2–12.** A floor plan is actually a section view of the building. (a) An imaginary cut is made at a level that passes through all windows and doors. (b) The floor plan shows what is left when the top is removed.

## ✓ CHECK YOUR PROGRESS

Can you perform these tasks?

- Identify oblique drawings.
- Identify isometric drawings.
- Identify orthographic drawings.
- Identify plan views.
- Identify elevation views.
- Identify section views.

## ASSIGNMENT

1. Identify each of the drawings in **Figure 2–13** as oblique, isometric, or orthographic.

2. Identify each of the drawings in **Figure 2–14** as elevation, plan, or section.

3. In the view of the house shown in **Figure 2–15**, which lines are true length?

4. What type of pictorial drawing is easiest to draw on the job site?

5. What type of drawing is used for working drawings?

(a)

(b)

**Figure 2–13.**

(b)

(c)

(a)

**Figure 2–14.**

**Figure 2–15.**

# Scales

**UNIT 3**

## Scale Drawings

Because construction projects are too large to be drawn full size on a sheet of paper, everything must be drawn proportionately smaller than it really is. For example, **floor plans** for a house are frequently drawn 1/48 of the actual size. This is called *drawing* to scale. At a scale of ¼" = 1'-0", ¼ inch on the drawing represents 1 foot on the actual building. When it is necessary to fit a large object on a drawing, a small scale is used. Smaller objects and drawings that must show more detail are drawn to a larger scale. The floor plan in **Figure 3–1** was drawn to a scale of ¼" = 1'-0". The detail drawing in **Figure 3–2** was drawn to a scale of 3" = 1'-0" to show the construction of one of the walls on the floor plan.

The scale to which a drawing is made is noted on the drawing. The scale is usually indicated alongside or beneath the title of the view.

## Reading an Architect's Scale

All necessary dimensions should be shown on the drawings. The instrument used to make drawings to scale is called an **architect's scale**, **Figure 3–3**. Measuring a drawing with an architect's or engineer's scale is a poor practice. At small scales it is especially difficult to be precise. The following discussion of how to read an architect's scale is presented only to ensure an understanding of the scales used on drawings. The triangular architect's scale includes eleven scales frequently used on drawings.

**Full Scale**

| | | | | | |
|---|---|---|---|---|---|
| 3/32" | = | 1' - 0" | 3/16" | = | 1' - 0" |
| 1/8" | = | 1' - 0" | 1/4" | = | 1' - 0" |
| 3/8" | = | 1' - 0" | 3/4" | = | 1' - 0" |
| 1/2" | = | 1' - 0" | 1" | = | 1' - 0" |
| 11/2" | = | 1' - 0" | 3" | = | 1' - 0" |

Two scales are combined on each face, except for the full-size scale, which is fully divided into sixteenths, **Figure 3–4.** The combined scales work together because one is twice as large as the other, and their zero points and extra divided units are on opposite ends of the scale.

The fraction, or number, near the zero at each end of the scale indicates the unit length in inches that is used on the drawing to represent 1 foot of the actual building. The extra unit near the zero end of the scale is subdivided into

## Objectives

After completing this unit, you will be able to perform the following tasks:

○ Identify the scale used on a construction drawing.

○ Read an architect's scale.

**Figure 3–1.** Portion of a plan view with a firewall.

In Figure 3-1 plan view:

- 36'-0"  36'-0"
- 7'-11 5/8"  9'-1 3/4"  7'-11 5/8"
- 5 1/2"  8'-8"  3 1/2"  8'-8 1/4"  5 1/2"  8'-8 1/4"  3 1/2"  8'-8"  5 1/2"
- BRM 2/0 / 2'-1"
- LINTEL: 2-2"x10's
- KITCHEN
- RF
- DW
- S / SNK
- RNG
- 3 1/2"
- 2'-8"
- 3'-6"
- DINING
- Area shown in Figure 3-2 →
- 2"x4"'s @ 8" O.C. STAGGERED 5/8" FIRE LODE GYP. BD. EA. SIDE
- ← FIREWALL
- 5 1/2" 4'-0"
- 4'-0"
- 4
- 5/8" F.C. GYP. BD. EA. SIDE
- 5 1/2"
- STORAGE
- 5'-5 1/8"
- 5 1/2"
- 7
- DINING
- KITCHEN
- RF
- DW
- RNG
- SNK / S
- 3'-6"
- 2'-8"
- 3 1/2" 3'-1"
- BRM 2/0 / 2'-1"
- 2
- TUB/
- WC
- L
- LIVING
- 5/8" F.C. GYP. BD. EA. SIDE
- 5'-5 1/8"
- STORAGE
- 4
- LIVING
- 7
- Elec. Panel
- 4 1/2"
- LINTEL: 2-2"x12's
- 4" CONC. PAD
- 17'-0 1/4"
- 3 1/2" 2'-1" 5 1/2" 5'-4" 3 1/2"
- ROOF OVERHANG
- 4" CONC. PAD
- 17'-0 1/4"
- LINTEL: 2-2"x12's
- 3 1/2"

**Figure 3–2.** Detail (plan at firewall).

- 5/8" FR GYP. BD. EACH SIDE
- 2X4 @ 16" O.C.
- 2X6 PLATES & SILL
- 8" (TYP)
- 6" BATT INSUL

**Figure 3–3.** Architect's scale. *Courtesy of Alvin & Co., Inc.*

**Figure 3–4.** Architect's triangular scales.

twelfths of a foot (inches) as well as fractions of inches on the larger scales.

To read the architect's scale, turn it to the ¼-inch scale. The scale is divided on the left from the zero toward the ¼ mark so that each line represents 1 inch. Counting the marks from the zero toward the ¼ mark, there are twelve lines marked on the scale. Each one of these lines is 1 inch on the ¼" = 1'-0" scale.

The fraction 1/8 is on the opposite end of the same scale. This is the 1/8-inch scale and is read in the opposite direction. Notice that the divided unit is only half as large as the one on the ¼-inch end of the scale. Counting the lines from zero toward the 1/8 mark, there are only six lines. This means that each line represents 2 inches at the 1/8-inch scale.

Now look at the 1½-inch scale. The divided unit is broken into twelfths of a foot (inches) and also fractional parts of an inch. Reading from the zero toward the number 1½, notice the figures 3, 6, and 9. These figures represent the measurements of 3 inches, 6 inches, and 9 inches at the 1½" = 1'-0" scale. From the zero to the first long mark, that represents 1 inch (which is the same length as the mark shown at 3) and four lines. This means that each line on the scale is equal to ¼ of an inch. Reading from the zero to the 3, read each line as follows: ¼, ½, ¾, 1, 1¼, 1½, 1¾, 2, 2¼, 2½, 2¾, and 3 inches. Do not confuse the engineer's scale with the architect's scale. The engineer's scale uses feet and decimal parts of a foot.

**Figure 3–5.**

## ✓ CHECK YOUR PROGRESS

Can you perform these tasks?

● Locate the scale notations on drawings.

● Use an architect's scale to measure objects drawn to scale.

## ASSIGNMENT

1. What are the dimensions indicated on the scale in **Figure 3–5?**

2. What scales are used for the following views of the Duplex? (Refer to the Duplex drawings in the packet.)
   a. Floor plan
   b. Site plan
   c. Front elevation
   d. Typical wall section

# Alphabet of Lines

The fact that drawings are used in construction for the communication of information was discussed earlier. The drawings, then, serve as a language for the construction industry. The basis for any language is its alphabet. The English language uses an alphabet made up of twenty-six letters. Construction drawings use an *alphabet of lines,* **Figure 4–1.**

The weight or thickness of lines is sometimes varied to show their relative importance. For example, in **Figure 4–2** notice that the basic outline of the building is heavier than the windows and doors. This difference in line weight sometimes helps distinguish the basic shape of an object from surface details.

## Object Lines

*Object lines* are used to show the shape of an object. All visible edges are represented by object lines. All the lines in **Figure 4–2** are object lines. Drawings usually include many solid lines that are not object lines, however. Some of these other solid lines are discussed here. Others are discussed later.

## Objectives

After completing this unit, you will be able to identify and understand the meaning of the listed lines:

- O  Object lines
- O  Dashed lines (hidden and phantom)
- O  Extension lines and dimension lines
- O  Centerlines
- O  Leaders
- O  Cutting-plane lines

(a) ——————————— HEAVY SOLID (FOR OBJECT LINES)

(b) ——————————— LIGHT SOLID (FOR DIMENSIONS, LEADER, AND EXTENSION LINES)

(c) – – – – – – – – – DASHED (FOR HIDDEN AND PHANTON LINES)

(d) —— – —— – —— – —— DOT DASH (FOR CENTERLINES)

(e) —— – – —— – – —— DOT DOT DASH (FOR CUTTING PLANES)

(f) —— ⁁ —— ⁁ —— BREAK LINES

**Figure 4–1.** Alphabet of lines

*Courtesy of Robert C. Kurzon*

**Figure 4–2.** Elevation outlined. *Courtesy of Robert C. Kurzon.*

**Figure 4–3.** The dashed lines on this foundation plan indicate the footing.

## Dashed Lines

Dashed lines have more than one purpose in construction drawings. One type of dashed line, the *hidden line,* is used to show the edges of objects that would not otherwise be visible in the view shown. Hidden lines are drawn as a series of evenly sized short dashes, **Figure 4–3.** If a construction drawing were to include hidden lines for all concealed edges, the drawing would be cluttered and hard to read. Therefore, only the most important features are shown by hidden lines.

Another type of dashed line is used to show important overhead construction, **Figure 4–4.** These dashed lines are called *phantom lines.* The objects they show are not hidden in the view—they are simply not in the view. For example, the most practical way to show exposed beams on a living room ceiling may be to show them on the floor plan with phantom lines. Phantom lines are also used to show alternate

DASHED ROOF EDGE

**Figure 4–4.** The dashed lines on this floor plan indicate the edge of the roof overhang.

positions of objects, **Figure 4–5.** To avoid confusion, the dashed lines may be made up of different weights and different length dashes, depending on the purpose, **Figure 4–6.**

# Extension Lines and Dimension Lines

*Extension lines* are thin, solid lines that project from an object to show the extent or limits of a dimension. Extension lines do not quite touch the object they indicate, **Figure 4–7.**

*Dimension lines* are solid lines of the same weight as extension lines. A dimension line is drawn from one extension line to the next. The dimension (distance between the extension lines) is lettered above the dimension line. On construction drawings, dimensions are expressed in feet and inches. The ends of dimension lines are drawn in one of three ways, as shown in **Figure 4–8.**

**Figure 4–5.** The dashed lines here are phantom lines to show alternate positions of the double-acting door and the door of the dishwasher.

**Figure 4–7.** Dimension and extension lines.

**Figure 4–6.** Different types of dashed lines are used to show different features.

**Figure 4–8.** Dimension line ends.

**Figure 4–10.** When centerlines show the center of a round object, the short dashes of two centerlines cross.

**Figure 4–9.** This centerline indicates that the column is symmetrical, or the same, on both sides of the centerline.

Dimensions that can be added together to come up with one overall dimension are called *chain dimensions.* The dimension lines for chain dimensions are kept in line as much as possible. This makes it easier to find the dimensions that must be added to find the overall dimension.

## Centerlines

**Centerlines** are made up of long and short dashes. They are used to show the centers of round or cylindrical objects. Centerlines are also used to indicate that an object is *symmetrical,* or the same on both sides of the center, **Figure 4–9.** To show the center of a round object, two centerlines are used so that the short dashes cross in the center, **Figure 4–10.**

SIDEWALK CORNER DETAIL

**Figure 4–11.** Method of showing the radius of an arc.

To lay out an arc or part of a circle, the radius must be known. The *radius* of an arc is the distance from the center to the edge of the arc. On construction drawings, the center of an arc is shown by crossing centerlines. The radius is dimensioned on a thin line from the center to the edge of the arc, **Figure 4–11.**

Rather than clutter the drawing with unnecessary lines, only the short, crossing dashes of the centerlines are shown. If the centerlines are needed to dimension the location of the center, only the needed centerlines are extended.

Figure 4–12. Leaders used for dimensioning.

SECTION VIEW

Figure 4–13. A cutting-plane line indicates where the imaginary cut is made and how it is viewed.

## Leaders

Some construction details are too small to allow enough room for clear dimensioning by the methods described earlier. To overcome this problem, the dimension is shown in a clear area of the drawing. A thin line called a *leader* shows where the dimension belongs, **Figure 4–12.**

## Cutting-plane Lines

It was established earlier that section views are needed to show interior detail. In order to show where the imaginary cut was made, a *cutting-plane line* is drawn on the view through which the cut was made, **Figure 4–13.** A cutting-plane line is usually a heavy line with long dashes and pairs of short dashes. Some

drafters, however, use a solid, heavy line. In either case, cutting-plane lines always have some identification at their ends and arrowheads to indicate the direction from which the section is viewed. Cutting-plane-line identification symbols are discussed in the next unit.

Some section views may not be referenced by a cutting-plane line on any other view. These are *typical sections* that would be the same if drawn from an imaginary cut in any part of the building, **Figure 4–14.**

ROOF
235# COMPOSITION SHINGLES
15# FELT
1/2" CDX PLYWD.
N.Y.S. APPROVED TRUSS @ 2'-0"
9" FIBERGLASS INSUL. W/V.B. R=33
1/2" GYP. BD.

MTL. DRIP EDGE
VENTED SOFFIT

TOP OF TRUSS

12
4

TOP OF PLATE

1'-0"

4'-6"

1'-0"

TYP. EXT. WALL
VINYL SIDING
1" STYROFOAM INSUL. R=5.5
2"x4"S @ 16" O.C.
3 1/2" FIBERGLASS INSUL. W/V.B. R=11
1/2" GYP. BD.

TYP. INT. WALL
2"x4"s @ 16" O.C.
1/2" GYP. BD. EA. SIDE

4" CONC. SLAB W/
6x6 10/10 WWM ON
POLY LOM MENBRANE ON
4" GRAVEL FILL

7'-11 1/2"

EXP. JT. FILLER

FIN. GRADE

8" CONC.

1" x 2'-0" POLYSTYRENE INSUL. R=5.5
SILL INSULATION

FIRST FLOOR
TOP OF FDN.
FIN. GRADE

5"

CONC. FOOTING 8"x16"
W/ 2-#4 BARS

3'-4.6" MIN.

TOP OF FTG.
BOT. OF FTG.

5"

TYPICAL SECTION
SCALE = 1/4" = 1'-0"

**Figure 4–14.** Building section.

## ✓ CHECK YOUR PROGRESS

Can you perform these tasks?

- Identify and explain the use of object lines.
- Identify and explain the use of hidden lines.
- Identify and explain the use of phantom lines.
- Identify and explain the use of dimension and extension lines.
- Identify and explain the use of centerlines.
- Identify and explain the use of leaders.
- Identify and explain the use of cutting-plane lines.

## ASSIGNMENT

Refer to the drawings of the Two-Unit Apartment in the packet. For each of the lines numbered A5.1 through A5.10, identify the kind of line and briefly describe its purpose on these drawings. The broad arrows with A5 numbers are for use in this assignment.

**Example:** A5.E, object line, shows end of building.

# Use of Symbols

An alphabet of lines allows for clear communication through drawings; the use of standard symbols makes for even better communication. Many features of construction cannot be drawn exactly as they appear on the building. Therefore, standard symbols are used to show various materials, plumbing fixtures and fittings, electrical devices, windows, doors, and other common objects. Notes are added to drawings to give additional explanations.

It is not important to memorize all the symbols and abbreviations used in construction before you learn to read drawings. You should, however, memorize a few of the most common symbols and abbreviations so that you may learn the principles involved in their use. Additional symbols and abbreviations can be looked up as they are needed. The illustrations shown here represent only a few of the more common symbols and abbreviations. A more complete reference is given in the Appendix.

## Door and Window Symbols

Door and window symbols show the type of door or window used and the direction the door or window opens. There are three basic ways for household doors to open—swing, slide, or fold, **Figure 5–1.** Within each of these basic types

## Objectives

After completing this unit, you will be able to identify and understand the meaning of the listed symbols:

○ Door and window symbols

○ Materials symbols

○ Electrical and mechanical symbols

○ Reference marks for coordinating drawings

○ Abbreviations

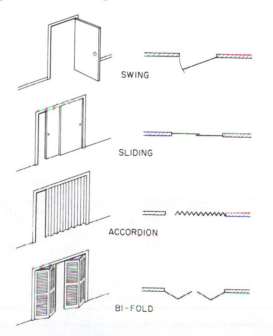

SWING

SLIDING

ACCORDION

BI-FOLD

**Figure 5–1.** Types of doors and their plan symbols.

PLAN

ELEVATION

DOUBLE HUNG

SLIDING

AWNING

CASEMENT

HOPPER

JALOUSIE

FIXED

**Figure 5–2.** Window symbols.

there are variations that can be readily understood from their symbols. The direction a swing-type door opens is shown by an arc representing the path of the door.

There are seven basic types of windows. They are named according to how they open, **Figure 5–2.** The symbols for hinged windows—**awning, casement,** and **hopper**—indicate the direction they open. In elevation, the symbols include dashed lines that come to a point at the hinged side, as viewed from the exterior.

The sizes of windows and doors are usually shown on a special window schedule or door schedule, but they might also be indicated by notes on the plans near their symbols. Door and window schedules are explained later. The notations of size show width first and height second. Manufacturers' catalogs usually list several sets of dimensions for every window model, **Figure 5–3.** The glass size indicates the area that will actually allow light to pass. The rough opening size is important for the carpenter who will frame the wall into which the window will be installed. The masonry opening is important to masons. The notations on plans and schedules usually indicate nominal dimensions. A *nominal dimension* is an approximate size and may not represent any of the actual dimensions of the unit. Nominal dimensions are usually rounded off to whole inches or feet and inches and are used only as a convenient way to refer to the window or door size. The actual dimensions should be obtained from the manufacturer before construction begins.

## Material Symbols

The drawing of an object shows its shape and location. The outline of the drawing may be filled in with a material symbol to show what the object is made of, **Figure 5–4.** Many materials are represented by one symbol in elevations and another symbol in sections. Examples of such symbols are concrete block and brick. Other materials look pretty much the same when viewed from any direction, so their symbols are drawn the same in sections and elevations.

When a large area is made up of one material, it is common to only draw the symbol in a part of the area, **Figure 5–5.** Some drafters simplify this even further by using a note to indicate what material is used and omitting the symbol altogether.

## Electrical and Mechanical Symbols

The electrical and mechanical systems in a building include wiring, electrical devices, piping, pipe fittings, plumbing fixtures, registers, and heating and air conditioning ducts. It is not practical to draw these items as they would actually appear, so standard symbols have been devised to indicate them.

The electrical system in a house includes wiring as well as devices such as switches, receptacles, light fixtures, and appliances. Wiring is indicated by lines that show how devices are connected. These lines are not shown in their actual position. They simply indicate which switches control which lights, for example. Outlets (receptacles) and switches are usually shown in their approximate positions. Major fixtures and appliances are shown in their actual positions. A few of the most common electrical symbols are shown in **Figure 5–6.**

Mechanical systems—plumbing and HVAC (heating, ventilating, and air conditioning)—are not usually shown in much detail on drawings for single-family homes. However, some of the most important features

**Figure 5–3.** Windows and doors can be measured in several ways.

## PLAN AND SECTION SYMBOLS

**Figure 5–4.** Material symbols.

**Figure 5–5.** Only part of the area is covered by the brick symbol, although the entire building will be brick.

may be shown. Piping is shown by lines; different types of lines represent different kinds of piping. Symbols for pipe fittings are the same basic shape as the fittings they represent. A short line, or *hash mark,* represents the joint between the pipe and the fitting. Plumbing fixtures are drawn pretty much as the actual fixture appears. A few plumbing symbols are shown in **Figure 5–7.**

## Reference Marks

A set of drawings for a complex building may include several sheets of section and detail drawings. These sections and details do not have much meaning without some way of knowing what part of the building they are meant to show.

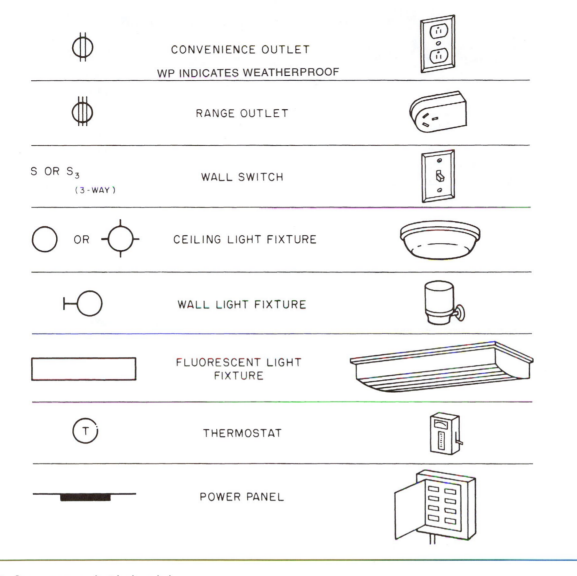

| | |
|---|---|
| $\phi$ | CONVENIENCE OUTLET<br>WP INDICATES WEATHERPROOF |
| $\phi$ | RANGE OUTLET |
| S OR $S_3$<br>(3-WAY) | WALL SWITCH |
| ◯ OR ⊕ | CEILING LIGHT FIXTURE |
| ⊢◯ | WALL LIGHT FIXTURE |
| ▭ | FLUORESCENT LIGHT FIXTURE |
| Ⓣ | THERMOSTAT |
| ▃ | POWER PANEL |

**Figure 5–6.** Some common electrical symbols.

Callouts, called *reference marks,* on plans and elevations indicate where details or sections of important features have been drawn. To be able to use these reference marks for coordinating drawings, you must first understand the numbering system used on the drawings. The simplest numbering system for drawings consists of numbering the drawing sheets and naming each of the views. For example, Sheet 1 might include a site plan and foundation plan; Sheet 2, floor plans; and Sheet 3, elevations.

On large, complex sets of drawings, the sheets are numbered according to the kind of drawings shown. Architectural drawing sheets are numbered A-1, A-2, and so on for all the sheets. Electrical drawings are numbered E-1, E-2, and E-3. A view number identifies each separate drawing or view on the sheet. **Figure 5–8** shows drawing 5 on Sheet A-4.

Because most of the drawings for a house are architectural, and the drawing set is fairly small, letters indicating the type of drawing are not usually included. Instead, the views are numbered, and a second number shows on which sheet it appears. For example, the fourth drawing on the third sheet would be 4/3, 4.3, or 4-3.

Numbering each view and the sheet on which it appears makes it easy to reference a section or detail to another drawing. The identification of a section view is given with the cutting-plane line showing where it is taken from. For example, the section view shown in **Figure 5–9** shows the fireplace at the cutting-plane line in **Figure 5–10**. Notice that the cutting-plane line in **Figure 5–10** indicates that the section is viewed from the top of the page toward the bottom, with the

VALVE

90° ELBOW

45° ELBOW

WYE

TEE

UNION

SOIL WASTE (BELOW GRADE)

SOIL WASTE (ABOVE GRADE)

COLD WATER

HOT WATER

G ———— G ———— GAS

PIPING SYMBOLS

**Figure 5–7.** Some common plumbing symbols.

fireplace opening on the right. That is how the section view in **Figure 5–9** is drawn. This numbering system is also used for details that cannot be located by a cutting-plane line. The drawing in **Figure 5–11** is a *typical wall section,* meaning that it is typical of a section view of any outside wall. This typical wall section includes a callout referencing a detail drawing of the roof cornice or eave. The cornice detail is Drawing 4 on Sheet A-4 and is shown in **Figure 5–12**. Notice that the detail faces the opposite direction from the typical wall section. That is because, being a typical section, not a specific section, it represents all exterior walls regardless of the direction they face.

## Abbreviations

Drawings for construction include many notes and labels of parts. These notes and labels are usually abbreviated as much as possible to avoid crowding the drawing. The abbreviations used on drawings are usually a shortened form of the word and are easily understood. For example, BLDG stands for building. The abbreviations used throughout this textbook and on the related drawings are defined in the Appendix.

5/4 PINE

3/4 PINE

1/2" QT. ROUND

FIXED SASH

5/A4 SECTION THRU 10" SIDELITE

SCALE 3"=1'-0"

**Figure 5–8.** This is drawing 5 on Sheet A-4.

12"X12" TC FLUE

2 X 4 WD. FRAMING

DAMPER

1" AIR SPACE

FIREBRICK

2- 2 X 10

#7 REINF. @ 1'-0"

CONC. FTG.

1" AIR SPACE

3/4" AD PLYWD.

1 X 8 PINE

3 1/2" MLD.

2 3/4" MLD.

BRICK

3 1/2"X 3 1/2"X 3/16" L

ASH DUMP

QUARRY TILE HEARTH

2- 2 X 10

PLYWD. FORM

CONC.

CONC. BLK.

C.O.

2.5 FIREPLACE SECTION
SCALE: 3/8"=1'-0"

**Figure 5–9.** This section view is drawing 2 on Sheet 5.

**Figure 5–10.** Plan for fireplace detailed in Figure 5–9.

**Figure 5–11.** The detail of this cornice is shown in drawing 4 on Sheet A-4, Figure 5–12.

SHINGLES AS SPEC.

$\frac{5}{8}$" PLYWOOD

METAL TRUSS

6" METAL CHANNEL
SUB-FASCIA

ALUM. DRIP EDGE

BLOCKING

PINE MOULDING #8013

1 X 8 WD. FASCIA

2 X 6 SUB-FASCIA

VENTED ALUM. SOFFIT

BOTTOM CHORD

6"

1$\frac{3}{4}$"

2$\frac{1}{2}$"

$\frac{3}{4}$"

24" O.H.

FASCIA DETAIL

SCALE : $\frac{1}{4}$ FULL SIZE

4
A-4

**Figure 5–12.** This is the detail of the cornice in Figure 5–11.

## ✓ CHECK YOUR PROGRESS

Can you perform these tasks?

- Identify window types by their symbols.

- Identify materials by their symbols.

- Identify the most common electrical equipment by its symbols.

- Identify the most common plumbing equipment by its symbols.

- Reference details by their symbols.

- Define several common abbreviations used on construction drawings.

## ASSIGNMENT

1. What is represented by each of these symbols?

   a.

   b.

   c.

   d.

   e.

   f.

   g. ——G——

   h. S₃

   i. WH

   j. WP

2. What is meant by each of these abbreviations?
   a. GYP. BD.
   b. FOUND.
   c. FIN. FL.
   d. O.C.
   e. REINF.
   f. EXT.
   g. COL.
   h. DIA.
   i. ELEV.
   j. CONC.

3. Where in a set of drawings would you find a detail numbered 6.4?

4. Where in a set of drawings would you find a detail numbered 5/M–3?

# Plan Views

You learned earlier in Unit 2 that plans are drawings that show an object as viewed from above. Many of the detail and section drawings in a set show parts of the building from above. Some of the plan views that show an entire building are discussed here. This brief explanation will help you feel more comfortable with plans, although it does not cover plans in depth. You will use plans frequently throughout your study of the remainder of this textbook. Each of the remaining units helps you understand plan views more thoroughly.

## Site Plans

A **site plan** gives information about the site on which the building is to be constructed. The boundaries of the site (property lines) are shown. The property line is usually a heavy line with one or two short dashes between longer line segments. The lengths of the boundaries are noted next to the line symbol. Property descriptions are usually the result of a survey by a surveyor. Surveyors and engineers usually work with decimal parts of feet, rather than feet and inches. Therefore, site dimensions are usually stated in tenths or hundredths of feet, **Figure 6–1.**

A symbol or north arrow of some type indicates what compass direction the site faces. Unless this north arrow includes a correction for the difference

## Objectives

After completing this unit, you will be able to explain the general kinds of information shown on the listed plans:

○ Site plans

○ Foundation plans

○ Floor plans

**Figure 6–1.** Minimum information shown on a site plan.

between true north and magnetic north, it may be only an approximation. However, it is sufficient to show the general direction the site faces.

The site plan also indicates where the building is positioned on the site. At a minimum, the dimensions to the front and one side of the site are given. The overall dimensions of the building are also included. Anyone reading the site plan will have this basic information without having to refer to the other drawings. If the finished site is to include walks, drives, or patios, these are also described by their overall dimensions.

## Foundation Plans

A foundation plan is like a floor plan, but instead of showing the living spaces, it shows the foundation walls and any other structural work to be done below the living spaces.

Two types of foundations are commonly used in homes and other small buildings. One type has a concrete base, called the **footing,** supporting foundation walls, **Figure 6–2.** The other is the slab-on-grade type. A *slab-on-grade* foundation consists of a concrete slab placed directly on the soil with little or no other support. Slabs-on-grade are usually thickened at their edges and wherever they must support a heavy load, **Figure 6–3.**

When the footing-and-wall-type foundation is used, girders are used to provide intermediate support to the structure above, **Figure 6–4.** The girder is shown on the foundation plan by phantom lines and a note describing it.

**Figure 6–2.** Footing and foundation wall.

**Figure 6–3.** Slab-on-grade foundation.

**Figure 6–4.** A girder provides intermediate support between the foundation walls.

FOUNDATION PLAN

*Courtesy of Home Planners, Inc.*

**Figure 6–5.** Foundation plan. *Courtesy of Home Planners, Inc.*

The foundation plan includes all the dimensions necessary to lay out the footings and foundation walls. The footings follow the walls and may be shown on the plan. If they are shown, it is usually by means of hidden lines to show their outline only. In addition to the layout of the foundation walls, dimensions are given for opening windows, doors, and ventilators. Notes on the plan indicate areas that are not to be excavated, concrete-slab floors, and other important information about the foundation, **Figure 6–5.**

## Floor Plans

A floor plan is similar to a foundation plan. It is a section view taken at a height that shows the placement of walls, windows, doors, cabinets, and other important features. A separate floor plan is included for each floor of the building. The floor plans provide more information about the building than any of the other drawings.

## Building Layout

The floor plans show the locations of all the walls, doors, and windows. Therefore, the floor plans show how the building is divided into rooms and how to get from one room to another. Before attempting to read any of the specific information on the floor plans, it is wise to familiarize yourself with the general layout of the building.

To quickly familiarize yourself with a floor plan, imagine that you are walking through the house. For example, imagine yourself standing in the front door of the left side of the Duplex—plans for which are included in the drawing packed with this text. You are looking across the living room. There is a closet on your right and a large window on your left. Straight ahead is the dining room with doors into a storage room and the kitchen. Looking in the kitchen doorway (notice there is no door in this doorway), you see cabinets, a sink, and a refrigerator on the opposite wall. More cabinets and a range are located on the left. Now, walk out of the kitchen and into the bedroom area. There are three doors; one leads into a large front bedroom with a long closet, another opens into a smaller bedroom, and the third opens into the bathroom. The bathroom includes a linen closet with bifold doors.

## Dimensions

Dimensions are given for the sizes and locations of all walls, partitions, doors, windows, and other important features. On frame construction, exterior walls are usually dimensioned to the outside face of the wall framing. If the walls are to be covered with stucco or masonry veneer, this material is outside the dimensioned face of the wall frame. Interior partitions may be dimensioned to their centerlines or to the face of the studs. (**Studs** are the vertical members in a wall frame.) Windows and doors may be dimensioned about their centerlines, **Figure 6–6,** or to the edges of the openings.

Solid masonry construction is dimensioned entirely to the face of the masonry, **Figure 6–7.** Masonry openings

**Figure 6–6.** Frame construction dimensioning.

**Figure 6–7.** Masonry construction dimensioning.

for doors and windows are dimensioned to the edge of the openings.

## Other Features of Floor Plans

The floor plan includes as much information as possible without making it cluttered and hard to read. Doors and windows are shown by their symbols, as explained in Unit 5. Cabinets are shown in their proper positions. The cabinets are explained further by cabinet elevations and details, which are discussed in Unit 8. If the building includes stairs, these are shown on the floor plan. Important overhead construction is also indicated on the floor plans. If the ceiling is framed with joists, their size, direction, and spacing are shown on the floor plan. Architectural features such as exposed beams, arches in doorways, or unusual roof lines may be shown by phantom lines.

## ✓ CHECK YOUR PROGRESS

Can you perform these tasks?

● Describe property boundaries from a simple site plan.

● Tell which direction a site faces from the site plan.

● Describe the position of a building within the site using the dimensions on a site plan.

● List the overall dimensions of a foundation from the information on a foundation plan.

● Identify girders on a foundation plan.

● Describe the locations of windows shown on plans.

● Describe the arrangement of rooms shown on a simple, one-story floor plan.

● Describe other major features shown on plans.

## ASSIGNMENT

Refer to the drawings for the Two-Unit Apartment (which are included in the accompanying packet) to complete this assignment.

1.  In what direction does the Apartment face?

2.  What is the length and width of the Apartment site?

3.  How far is the front of the Apartment from the front property line?

4.  What is the overall length and width of the Apartment?

5.  What are the inside dimensions of the front bedroom?

6.  What is the thickness of the partitions between the two bedrooms?

7.  What is the thickness of the interior wall between the two dining rooms?

8.  With two exceptions, the units in the Apartment are exactly reversed. What are the two exceptions?

9.  What is the distance from the west end of the Apartment to the centerline of the west front entrance?

10. What is indicated by the small rectangle on the floor plan outside each main entrance?

11. What is the distance from the ends of the Apartment to the centerlines of the $6^0/6^8$ sliding glass doors?

12. What is indicated by the dashed line just outside the front and back walls on the floor plan of the Apartment?

# Elevations

## Objectives

After completing this unit, you will be able to perform the following tasks:

○ Orient building elevations to building plans.

○ Explain the kinds of information shown on elevations.

Drawings that show the height of objects are called *elevations.* However, when builders and architects refer to building elevations, they mean the exterior elevation drawings of the building, **Figure 7–1.** A set of working drawings usually includes an elevation of each of the four sides of the building. If the building is very complex, there may be more than four elevations. If the building is simple, there may be only two elevations—the front and one side.

FRONT ELEVATION

SIDE ELEVATION

REAR ELEVATION

**Figure 7–1.** Building elevations.

# Orienting Elevations

It is important to determine the relationship of one drawing to another. This is called *orienting* the drawings. For example, if you know which elevation is the front, you must be able to picture how it relates to the front of the floor plan.

Elevations are usually named according to compass directions, **Figure 7–2.** The side of the house that faces north is the north elevation, and the side that faces south is the south elevation, for example. When the elevations are named according to compass direction, they can be oriented to the floor plan, foundation plan, and site plan by the north arrow on those plans. It might help to label the edge of the plans according to the north arrow, **Figure 7–3.**

It is not always possible to label elevations according to compass direction, however. When drawings are prepared to be sold through a catalog or when they are for use on several sites, the compass directions cannot be included. In this case, the elevations are named according to their position as you face the building, **Figure 7–4.** To orient these elevations to the plans, find the front on the plans. The front is usually at the bottom of the sheet, but it can be checked by the location of the main entrance.

# Information on Building Elevations

Building elevations are normally quite simple. Although the elevations do not include a lot of detailed dimensions and notes, they show the finished appearance of the building better than other views. Therefore, elevations are a great aid in understanding the rest of the drawing set.

The elevations show most of the building, as it will actually appear, with solid lines. However, the underground portion of the foundation is shown as

**Figure 7–2.** Elevations are usually named according to their compass directions.

E

N

S

FLOOR PLAN

W

*Courtesy of Robert C. Kurzon*

**Figure 7–3.** Plan labeled to help orientation to north arrow. *Courtesy of Robert C. Kurzon.*

REAR          LEFT          FRONT          RIGHT

LEFT          REAR

FRONT          RIGHT

**Figure 7–4.** Elevations can be named according to their relative positions.

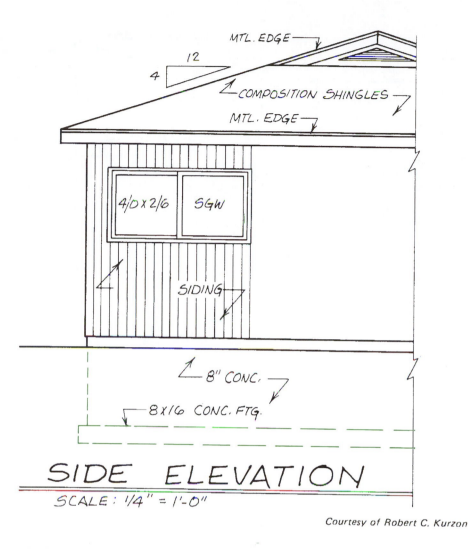

## SIDE ELEVATION
SCALE: 1/4" = 1'-0"

MTL. EDGE

12
4

COMPOSITION SHINGLES

MTL. EDGE

4/0 X 2/6   SGW

SIDING

8" CONC.

8X16 CONC. FTG.

*Courtesy of Robert C. Kurzon*

**Figure 7–5.** Underground portion of the building is shown with dashed lines. *Courtesy of Robert C. Kurzon.*

hidden lines, **Figure 7–5.** The footing is shown as a rectangle of dashed lines at the bottom of the foundation walls.

The surface of the ground is shown by a heavy solid line, called a grade line. The *grade line* might include one or more notes to indicate the elevation above sea level or another reference point, **Figure 7–6.** Elevation used in this sense is altitude, or height—not a type of drawing. All references to the height of the ground or the level of key parts of the building are in terms of elevation. Methods for measuring site elevations are discussed in Unit 9.

Some important dimensions are included on the building elevations. Most of them are given in a string

at the end of one or more elevations, **Figure 7–7.** The dimensions most often included are listed here:

○ Thickness of footing

○ Height of foundation walls

○ Top of foundation to finished first floor

○ Finished floor to ceiling or top of plate (The **plate** is the uppermost framing member in the wall.)

○ Finished floor to bottom of window headers (The **headers** are the framing across the top of a window opening.)

○ Roof overhang at eaves

**Figure 7–6.** The elevation of this site is 150 feet.

**Figure 7–7.** Dimensions on an elevation.

## ✓ CHECK YOUR PROGRESS

Can you perform these tasks?

- Explain which side of a plan view is represented by a building elevation.

- Identify footings and foundations on building elevations.

- Find the height of a foundation wall on the building elevations.

- Find the dimension from the floor to the top of a wall on building elevations.

- Find the amount of roof overhang shown on building elevations.

- Describe the appearance of a building from information given on the building elevations.

## ASSIGNMENT

Refer to the drawings of the Two-Unit Apartment in the packet to complete this assignment.

1. Which elevation is the north elevation?

2. In what compass direction does the left end of the Apartment face?

3. What is the dimension from the surface of the floor to the top of the wall framing?

4. What is the thickness of the floor?

5. How far does the foundation project above the ground?

6. How far below the surface of the ground does the foundation wall extend?

7. What is the total height of the foundation walls?

8. What is the minimum depth of the bottom of the footings?

# Sections and Details

It is not possible to show all the details of construction on foundation plans, floor plans, and building elevations. Those drawings are meant to show the relationships of the major building elements to one another. To show how individual pieces fit together, it is necessary to use larger-scale drawings and section views. These drawings are usually grouped together in the drawing set. They are referred to as **sections** and **details, Figure 8–1.**

## Sections

Nearly all sets of drawings include, at least, a typical wall section. The typical section may be a section view of one wall, or it may be a full section of the building. Full sections are named by the direction in which the imaginary cut is made. **Figure 8–2** shows a transverse section. A *transverse section* is taken from an imaginary cut across the width of the building. Transverse sections are sometimes called *cross sections.* A full section taken from a lengthwise cut through the building is called a *longitudinal section,* **Figure 8–3.**

Full sections and wall sections normally have only a few dimensions but have many notes with leaders to identify the parts of the wall. The following is a list of the kinds of information that are included on typical wall sections with most sets of drawings:

- Footing size and material (This is specified by building codes.)

- Foundation wall thickness, height, and material

- Insulation, waterproofing, and interior finish for foundation walls

- Fill and waterproofing under concrete floors

- Concrete floor thickness, material, and reinforcement

- Sizes of floor framing materials

- Sizes of wall framing materials

- Wall covering (sheathing, siding, stucco, masonry, and interior wall finish) and insulation

- Cornice construction—materials and sizes (The **cornice** is the construction at the roof eaves.)

- Ceiling construction and insulation

## Objectives

After completing this unit, you will be able to perform the following tasks:

○ Find and explain information shown on section views.

○ Find and explain information shown on large-scale details.

○ Orient sections and details to the other plans and elevations.

**Figure 8-1.** Typical sheet of sections and details for a small building.

ROOF
235# COMPOSITION SHINGLES
15# FELT
7/16" O.S.B. 24/16
N.Y.S. APPROVED TRUSS @ 2'-0"
9" FIBERGLASS INSUL. W/V.B. R=33
1/2" GYP. BD.

MTL. DRIP EDGE
VENTED SOFFIT

TYP. EXT. WALL
VINYL SIDING
1" STYROFOAM INSUL. R=5.5
7/16" O.S.B. 24/16
2" x 4's @ 16" O.C.
3 1/2" FIBERGLASS INSUL. R=11
VAP. B. BARRIER
1/2" GYP. BD.

FIN. GRADE

8" CONC. BLOCK

CONC. FOOTING 8"X16"
W/ 2-#4 BARS

TYP. INT. WALL
2"x4"s @ 16" O.C.
1/2" GYP. BD. EA. SIDE

EXP. JT. FILLER

1" X 2'-0" POLYSTYRENE INSUL. R=5.5
SILL INSULATION

4" CONC. SLAB W/
6x6 10/10 WWM ON
6 MIL POLY FILM MENBRANE ON
4" GRAVEL FILL

TOP OF TRUSS

12
4

TOP OF PLATE

FIRST FLOOR
TOP OF FDN.
FIN. GRADE

TOP OF FTG.
BOT. OF FTG.

## TYPICAL SECTION
SCALE = 1/4" = 1'-0"

**Figure 8–2.** Transverse section.

ROOF
APPROVED TRUSSES @ 2'-0" O.C.
#15 FELT ON 5/8" CDX
#235 SHINGLES

12
4

TYP. INT. WALL
2X4 @ 16" O.C.
1/2" GYP. BD. EA. SIDE

BLOCKING

FIREWALL
2X6 PLATES
2X4 STUDS
@ 8" O.C.
STAGGERED

GRADE LINE

8" CONC.

8"X16" CONC. FTG.

4" CONC. SLAB
W/6"X6" - #10 WWM
ON 6 MIL. POLY FILM AND
COMPACTED GRAVEL

**Figure 8–3.** Longitudinal section.

Other section drawings are included as necessary to explain special features of construction. Wherever wall construction varies from the typical wall section, another wall section should be included. Section views are used to show any special construction that cannot be shown on normal plans and elevations. **Figure 8–4** is an example of a special section in elevation. This section view is said to be in elevation because it shows the height of the ridge construction. **Figure 8–5** is in plan because it shows the interior of the fireplace as viewed from above.

**Figure 8–4.** Special section of ventilated ridge. *Courtesy of W. D. Farmer.*

**Figure 8–5.** A section in plan. *Courtesy of W. D. Farmer.*

## Other Large-Scale Details

Sometimes necessary information can be conveyed without showing the interior construction. A large scale may be all that is needed to show the necessary details. The most common examples of this are on cabinet installation drawings, **Figure 8–6.** Cabinet elevations show how the cabinets are located, without showing the interior construction.

Many details are best shown by combining elevations and sections or by using isometric drawings. **Figure 8–7** shows an example of an elevation and a section used together to explain the construction of a fireplace. **Figure 8–8** shows an isometric detail drawing that includes sections to show interior construction.

SOFFIT

W30x12   W15x30
         W15x30
              W15x30

EXHAUST HOOD

OPEN SPICE SHELVES

RANGE   4" BACK SPLASH

B15-D   B33-C

SOFFIT

W40x12          W30x12

        W30x30

4" BACK SPLASH

REF.

SF36   DW.

KITCHEN CABINETS
SCALE: 3/8" = 1'-0"

**Figure 8–6.** Cabinet elevations.

FOR WALL FINISH SEE SPEC.

CHOICE OF MANTEL BY OWNER

ELEVATION

2x4 @ 16" O.C.   FLUE

2-2 x 10 ON EDGE

SMOKE CHAMBER

1/2" MIN.

8"

ADJUST-ABLE DAMPER

THROAT 8"

SMOKE SHELF

FIRE BRICK

ASH DUMP

SECTION

**Figure 8–7.** Fireplace details. *Courtesy of W. D. Farmer.*

2X6 STUDS -24"O.C.

INTERIOR FINISH

VAPOR BARRIER

R-19 INSULATION (6" BATT)

1X6 OR BACK-UP CLIP

½" SHEATHING

**Figure 8–8.** Isometric section. *Courtesy of W. D. Farmer.*

## Orienting Sections and Details

As explained earlier, some sections and details are labeled as typical. These drawings describe the construction that is used throughout most of the building.

Sections and details that refer to a specific location in the building include a reference that indicates where the section or detail came from. That larger source drawing has a cutting-plane line to show exactly where the section cut or detail is taken from. The cutting-plane line has an arrowhead or some other indication of the direction from which the detail is viewed. The top drawing in **Figure 8–9** shows that there is a section view or detail of the construction at the skylight. The little flag at the top points to the right, so that is the direction from which

the detail at the bottom of the figure is viewed. The label on the bottom drawing, the skylight detail, includes the number of the drawing, corresponding to the number indicated at the cutting-plane line in the top view. A reference mark near the arrow indicates where the detail drawing is shown. The reference marks that are used for orienting details may vary from one set of drawings to another. It is important, although not usually difficult, to study the drawings and learn how the architect references details. Usually a system of sheet numbers and view numbers is used. One such numbering system was explained earlier.

Some basic principles of details and sections have been discussed here. You will gain more practice later in reading details and sections.

MTL. FLUE

COMP. SHINGLES

SKYLIGHT

WD. SIDING

## NORTH ELEVATION
1/4" = 1'-0"

DBL. DOME PLAS. SKYLITE

16 OZ. BASE FLASH (INTEGRAL)

COMP. SHINGLES

PLYWD. LINER (INTEGRAL)

1 X T.&G. PINE

1 X 2 PINE

## SKYLIGHT DETAIL
3" = 1'-0"

**Figure 8–9.** The elevation at the top indicates that the construction detail is drawing 2 on Sheet 7. The bottom drawing is that detail.

## ✓ CHECK YOUR PROGRESS

Can you perform these tasks?

● Find the part of a building from which a section view was drawn.

● Explain the notes commonly included on typical wall sections.

● Explain whether a section view is in elevation or in plan.

● Find the part of a building from which a large-scale detail was drawn.

## ASSIGNMENT

Refer to the drawings of the Two-Unit Apartment in the packet to complete the assignment.

1. What is used to show the detail of a complex design, installation, or product?

2. What kind of section drawing is the typical wall section on Sheet 2?

3. What kind and size material is to be used for the foundation walls?

4. What is used between the concrete-slab floor and the exterior wall framing?

5. What kind and size of insulation is used around the foundation? Is this insulation used on the inside or outside of the foundation?

6. What kind and size of material is to be used on the inside of the frame walls?

7. Sheet 2 includes a firewall detail. Where in the Apartment is this firewall?

8. What is the distance between the centerlines of the studs in the firewall?

9. What is the total thickness of the firewall? (Remember that a $2 \times 6$ is actually 5½" wide.)

10. Were the cabinet elevations drawn of the kitchen on the east side or the west side of the Apartment?

11. How would the kitchen elevations be different if they were drawn from the other kitchen?

12. What is the distance from the kitchen countertop to the bottom of the wall cabinets?

13. How far does the roof overhang project beyond the exterior walls?

14. Where are the electrical panels located?

# TEST

**A. Identify each of the dimensions indicated on the illustrated scale.**

**B. Which occupation or individual listed in Column II performs the task listed in Column I?**

| I | II |
|---|---|
| 1. Obtains a building permit | a. Architect |
| 2. Issues the building permit | b. Building inspector |
| 3. Acts as the owner's representative | c. Owner |
| 4. Issues certificate of occupancy | d. Mechanical engineer |

5. Lays out rooms for efficient use
6. Designs plumbing in large buildings
7. Hires and supervises carpenters
8. Checks to see that codes are observed

e. Municipal building department
f. General contractor

**C. Which of the lines shown in Column II is most likely to be used for each purpose in Column I?**

| I | | II |
|---|---|---|
| 1. Outline of a window | a. _____ |
| 2. Alternate position of a fold-down countertop | b. _____ |
| 3. Centerline of a round post | c. _____ |
| 4. Extension line to show extent of a dimension | d. _____ |
| 5. Buried footing | e. _____ |
| 6. Point at which an imaginary cut is made for a section view | f. _____ |

**D. Which of the symbols shown in Column II is used for each of the objects or materials in Column I?**

I

1. Awning window in elevation
2. Bifold door in plan
3. Earth
4. Rough wood
5. Batt insulation
6. Concrete
7. Ceiling light fixture
8. Finish wood
9. Shutoff valve (plumbing)
10. Hopper window in elevation

II

**E. Select the one best answer for each question.**

1. What kind of regulation controls the types of buildings allowed in each part of a community?
   a. building code
   b. zoning law
   c. specification
   d. certificate of occupancy

2. Which of the listed kinds of information can be clearly shown on construction drawings?
   a. size of parts
   b. location of parts
   c. shape of parts
   d. all of these

3. What type of drawing is the 2 × 4 shown in Illustration 3?
   a. isometric
   b. oblique
   c. perspective
   d. none of these

*oblique*

4. What type of drawing is the 2 × 4 shown in Illustration 4?
   a. isometric
   b. oblique
   c. perspective
   d. none of these

*isometric*

5. What type of drawing is the 2 × 4 shown in Illustration 5?
   a. isometric
   b. oblique
   c. perspective
   d. none of these

*NONE of the above*

6. What kind of drawing is shown in Illustration 6?
   a. elevation
   b. detail
   c. rendering
   d. plan

7. What kind of drawing is shown in Illustration 7?
   a. elevation
   b. detail
   c. rendering
   d. plan

8. What kind of drawing is shown in Illustration 8?
   a. elevation
   b. detail
   c. rendering
   d. plan

SECTION THRU 10" SIDELITE
SCALE 3" = 1'-0"

9. If an object 12 feet long is drawn at a scale of ¼"= 1'-0", how long is the drawing?
   a. 48 inches
   b. 3 inches
   c. 3 feet
   d. none of these

10. If an object 1'-6"long is drawn at a scale of 1½"= 1'-0", how long is the drawing?
    a. 2¼ inches
    b. 2 inches
    c. 27 inches
    d. none of these

11. Where in a set of drawings would you find detail number 9.6?
    a. sheet 6
    b. sheet 9
    c. ninth sheet in the mechanical section
    d. none of the above

12. In the drawing key 4/A–4, what does the letter A stand for?
    a. architect's initial
    b. first edition of the drawings
    c. architectural
    d. first detail on the sheet

13. On which drawing would you expect to find the height of the foundation wall?
    a. site plan
    b. building elevation
    c. floor plan
    d. foundation plan

14. On which drawing would you expect to find the setback of the building?
    a. site plan
    b. building elevation
    c. floor plan
    d. foundation plan

15. On which drawing would you expect to find the height of the window heads?
    a. site plan
    b. building elevation
    c. floor plan
    d. window detail

## F. Refer to the Two-Unit Apartment drawings to answer these questions.

1. How far is the building from the west boundary?

2. What is the dimension from the finished floor to the top of the wall plate?

3. What is the overall length of the building at window height?

4. What is the overall length of the building at the eaves?

5. What is the north-to-south dimension inside the front bedrooms?

6. What is the slope of the roof?

7. What types of windows are used?

8. How thick is the concrete footing?

9. What material is the foundation wall?

10. What is under the floor at its outer edges?

# READING DRAWINGS FOR TRADE INFORMATION

In Part II, you will examine all the information necessary to build a moderately complex single-family home. The sequence of the units in Part II follows the sequence of actual construction. In some cases, all the information necessary for a particular phase of construction can be found on one sheet of drawings. Other phases require cross-referencing among several drawings. The relationships among the various drawings are discussed as the need to cross-reference them arises.

The assignments in this part refer to the Lake House drawings provided in the packet. The Lake House was designed as a vacation home on a lake in Virginia. The design is moderately complex, involving several floor levels and some interesting construction techniques.

# UNIT 9

# Clearing and Rough Grading the Site

## Objectives

After completing this unit, you will be able to perform the following tasks:

○ Identify work to be included in clearing a building site according to site plans.

○ Interpret grading indications on a site plan.

○ Interpolate unspecified site elevations.

## Property Boundary Lines

The boundary lines of the building site are shown on the site plan. The direction of a property line is usually expressed as a bearing angle. The *bearing* of a line is the angle between the line and north or south. Bearing angles are measured from north or south depending on which keeps the bearing under 90°, **Figure 9–1.** Angles are expressed in degrees (°), minutes ('), and seconds ("). There are 360 degrees in a complete circle, 60 minutes in a degree, and 60 seconds in a minute.

The point of beginning (P.O.B.) may or may not be shown on the site plan. If the point of beginning is not shown on the plan, start at a convenient corner.

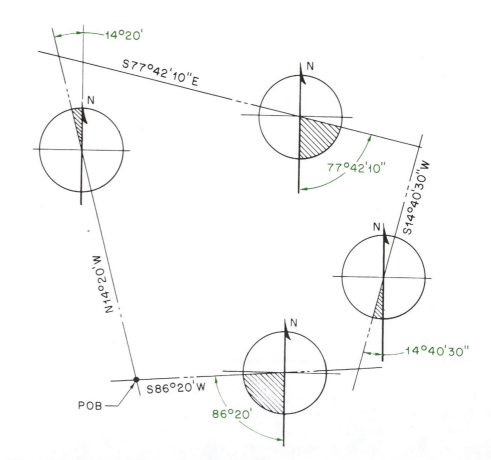

**Figure 9–1.** Bearing angles are always less than 90°.

Corners are usually marked with an iron pin (I.P.) or some permanent feature. The approximate direction of the boundaries can be found with a handheld compass. This approximation should be accurate enough to aid in finding the marker (iron pin, manhole cover, concrete marker, or similar item) at the next corner. Proceed around the perimeter in this manner to find all corners. All construction activity should be kept within the property boundaries unless permission is first obtained from neighboring landowners.

**Figure 9–2.** The winter sun passes through deciduous trees.

## Clearing the Site

The first step in actual construction is to prepare the site. This means clearing any brush or trees that are not to be part of the finished landscape. The architect's choice of trees to remain is based on consideration of many factors. Trees and other natural features can be an important part of architecture—not only for their natural beauty, but for energy conservation. For example, deciduous trees, which lose their leaves in the winter, can be used to effectively control the solar energy striking a house. In the winter, the sun shines through the deciduous trees on the south side of a house, **Figure 9–2.** In the summer, the trees shade the south side of the house, **Figure 9–3.** The Lake House offers a good example of the importance of the selection of trees to remain on a site. This house gets a large part of its heat from its passive-solar features. The passive-solar features are described more fully later.

    Trees that are to be saved are shown on the plot plan by a symbol and a note indicating their butt diameter and species, **Figure 9–4.** Areas that are too densely wooded to show individual trees are outlined and marked "woods," **Figure 9–5.** Removal of unwanted trees may require felling and stump removal, or may be accomplished with a bulldozer and dump truck. In either case, care must be exercised not to damage the trees that are to be saved.

**Figure 9–3.** The summer sun is shaded by deciduous trees.

**Figure 9–4.** Typical note and symbol for individual tree.

## Grading

*Grading* refers to moving earth away from high areas and into low areas. Site grading is necessary to ensure that water drains away from the building properly and does not puddle or run into the building. In some cases, grading may be necessary for access to the site. For example, if

**Figure 9–5.** Typical note and symbol for wooded area.

**Figure 9–6.** Spot elevations for specific locations.

the site has a steep grade, it may be necessary to provide a more gradual slope for a driveway.

Grade is measured in vertical feet from sea level or from a fixed object such as a manhole cover. This vertical distance is called *elevation.* The term *elevation* to denote a vertical position should not be confused with elevation drawings that show the height of objects. The elevations of specific points are given as *spot elevations.* Spot elevations are used to establish points in a driveway, a walk, or the slope of a terrace, **Figure 9–6.** Spot elevations are often given for trees that are to be saved.

The grade of a site is shown by *topographic* **contour lines.** These are lines following a particular elevation. The vertical difference between contour lines is the **vertical contour interval.** For plot plans this is usually 1 or 2 feet. When the land slopes steeply, the contour lines are closely spaced. When the slope is gradual, the contour lines are more widely spaced.

The builder must be concerned with not only the grade or contour of the existing site, but also that of the finished site. To show both contours, two sets of contour lines are included on the plot plan. Broken lines indicate natural grade (N.G.), and solid lines indicate finished grade (F.G.), **Figure 9–7.**

When the natural-grade elevation is higher than the finished-grade elevation, earth must be removed. This is

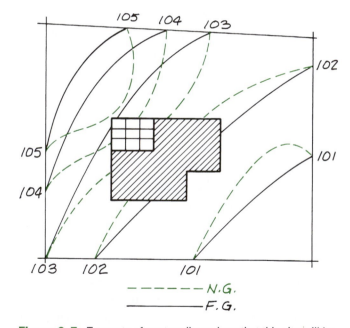

**Figure 9–7.** Two sets of contour lines show that this site will be graded to be more level in the area of the building.

referred to as *cut.* When the natural grade is at a lower elevation than the finished grade, *fill* is required. To determine the amount of cut or fill required at a given point, find the difference between the natural grade and the finished grade, **Figure 9–8.**

NATURAL GRADE

104'

103'

FIN. GRADE

102'

101'

100'

CUT

FILL

CUT

**Figure 9–8.** Cutting is required where N.G. is above F.G. Fill is required where N.G. is below F.G.

## Interpolating Elevations

Sometimes it is necessary to find an elevation that falls between two contour lines. This can be done by interpolation. *Interpolation* is a method of finding an unknown value by comparing it with known values.

**Example:** To interpolate the elevation of the tree at point A in **Figure 9–9,** follow the listed steps of the procedure using the information shown in the illustration and the numbers enclosed in parentheses.

Step 1.    Scale the distance between the two adjacent contour lines (12 feet).

Step 2.    Scale the distance from the unknown point to the nearest contour line (4 feet).

Step 3.    Multiply the contour interval by the fraction of the distance between the contour lines to the unknown point. (Contour interval = 2 feet; fraction of distance between contour lines = 4/12 = 1/3. Therefore, 2 = 1/3 = 2/3 feet.)

Step 4.    If the nearest contour line is below the other one, add this to it. If the nearest contour line is above the other one, subtract this amount. (Nearest contour = 48'. This is below the other contour line at 50', so 2/3' is added to 48'. 2/3' + 48' = 48.66'.)

48'

A—⊕—TREE

50'

SCALE: $\frac{1"}{8}$ = 1'-0"

**Figure 9–9.** Interpolate the elevation of the tree.

## ✓ CHECK YOUR PROGRESS

Can you perform these tasks?

- ● Determine the bearings and lengths of property lines.

- ● Identify each tree to be removed from a site.

- ● Give the natural grades and finished grades of all points on a site.

- ● Explain which direction water will naturally run from any point on a site.

## ASSIGNMENT

Refer to the site drawings of the Lake House (in the packet) to complete the assignment.

1. What are the lengths of the north, east, and west boundaries of the Lake House?

2. Which of the compass points shown in **Figure 9–10** corresponds with the north boundary of the Lake House? Which compass point corresponds with the east boundary?

3. How many trees are indicated for removal?

4. How many trees are to remain on the site? (Do not include wooded areas.)

5. What is the finished-grade elevation at the tree nearest the Lake House?

6. What is the natural-grade elevation of the most easterly tree to be saved?

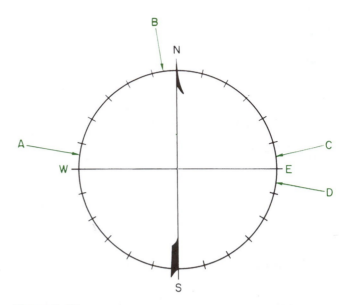

**Figure 9–10.**

7. What is the elevation of the tree to be saved nearest the lake?

8. What is the natural-grade elevation at the southwest corner of the Lake House? Do not include the deck as part of the house.

9. What is the finished-grade elevation at the south-west corner of the house?

10. How much cut or fill is required at the entrance of the garage?

11. Is cut or fill required at the southwest corner of the house? How much?

12. What is the elevation at the northeast corner of the site?

# Locating the Building

## Laying Out Building Lines

The position of the building is shown on the site plan. Dimensions show the distance from the street (or lake) to the building and from the side boundaries to the building. The location of one corner can easily be found by measuring with a long (100′ or 200′) steel tape. The most efficient way to find all remaining corners is by the use of a leveling instrument, **Figure 10–1.** The functions of the parts of a leveling instrument are as follows:

○ *Telescope* contains the lens, focusing adjustment, and cross hairs for sighting.

○ *Telescope level* is a spirit level used for leveling the instrument prior to use.

○ *Clamp screw* locks the instrument in position horizontally.

○ *Fine adjusting screw* makes fine adjustments in a horizontal plane.

○ *Leveling base* holds the leveling screws (usually four) for leveling the instrument prior to use.

○ *Protractor* is a scale graduated in degrees and minutes for measuring horizontal angles.

Two accessories are required for most operations performed with a leveling instrument. The tripod is a three-legged stand that provides a stable base for the instrument. A *target rod* is a separate device with a scale graduated in feet and tenths of a foot. The telescope is focused on the target rod to measure elevations. The builder's level is a device for checking the difference in elevation between

## Objectives

After completing this unit, you will be able to perform the following tasks:

○ Lay out building lines according to a site plan.

○ Use the 6-8-10 or equal-diagonals method to check the squareness of corners.

○ Use a leveling instrument to measure angles and depths of excavations.

**Figure 10–1.** Builder's level. *Courtesy of David White.*

TELESCOPE FOCUSING KNOB
TELESCOPE
PROTRACTOR
LEVELING BASE
SPIRIT LEVEL
CLAMP SCREW
LEVELING SCREWS

**Figure 10–2.** A transit is similar to a builder's level, but it can be tilted to measure vertical angles.

two points. It can also be used for measuring angles on a horizontal plane. Another instrument, a *transit,* can be tilted to measure angles in a vertical plane, **Figure 10–2.** The procedure described here for laying out square corners can be used with either instrument:

Step 1.   Set the tripod up over a known corner. The exact position is determined by hanging a plumb bob from the tripod. The legs should be firmly set in the ground about 3 feet apart.

Step 2.   Set the instrument on top of the tripod and hand tighten the clamp screw.

Step 3.   Turn the leveling screws down so they contact the tripod plate.

Step 4.   Turn the telescope so that it is over one pair of leveling screws. Adjust these two screws so that the telescope is level.

Step 5.   Rotate the telescope so that it is over the other pair of leveling screws. Adjust these two screws to level the telescope.

Step 6.   Repeat this over each pair of leveling screws until the telescope is level in all positions.

Step 7.   Using a compass, carefully point the telescope to magnetic north. The north arrow on the drawing can point to either true north or magnetic north. The compass needle will point only

to magnetic north. Because the true north pole of the earth is some distance from its magnetic north, the difference between bearings based on true north and those based on magnetic north can be several degrees. If the building is to be laid out according to true-north bearings, not magnetic compass bearings, contact the architect or surveyor to find out what correction should be made in your area.

Step 8.   Set the protractor at zero degrees.

Step 9.   Rotate the telescope to the bearing of one building line.

Step 10.   Stretch a line (string) from the existing stake to several feet beyond the next corner.

Step 11.   Have a partner hold the target rod plumb over the far end of this line. When the telescope cross hairs can be focused on the target rod, the bearing of the line is correct.

Step 12.   Measure the length of this building line from the first stake, and drive another stake.

Step 13.   To lay out each of the remaining corners, set the tripod over the stake, line the telescope up on the existing line, and then use the protractor of the instrument to measure a 90° corner.

To check the squareness of the layout, measure the diagonals of a rectangle formed by the layout, **Figure 10–3.** When all four corners of the rectangle are 90°, the diagonals are equal.

Another method of checking a 90° angle is called the *6-8-10 method,* **Figure 10–4.** Measure 6 feet from the corner along one line. Measure 8 feet from the corner along the other line. These points should be 10 feet apart. (See Math Review 24.)

**Figure 10–3.** When the diagonals of a rectangle are the same length, the corners are square.

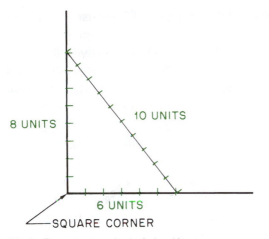

8 UNITS

10 UNITS

6 UNITS

SQUARE CORNER

**Figure 10–4.** The 6-8-10 method of checking a square corner.

BUILDING LINES
TRANSFERRED TO
BATTER BOARDS

BATTER
BOARDS

PLUMB BOB

BUILDING
LINES

**Figure 10–5.** Batter boards.

BENCHMARK
99.0'

GRADELINE AT TOP
OF EXCAVATION
101.3'

BOTTOM OF
EXCAVATION
94.8'

**Figure 10–6.** The top of this excavation is 2.3 feet above the benchmark. Its depth is 6.5 feet (101.3'-94.8').

The **building lines** can be saved, even after the corner stakes are removed for earthwork, by erecting batter boards, **Figure 10–5. Batter boards** are sturdy horizontal boards fastened between 2 × 4 stakes, at least 4 feet outside the building lines. The building lines are extended and marked on the batter boards.

## Excavating

Most buildings require some *excavation* (digging) to prepare the site for a foundation. The depth of the excavation is measured from a fixed benchmark. A benchmark can be any stationary object such as a surveyed point on a street or very large boulder. All elevations (vertical distances) are measured from this benchmark. Only in the case of a real coincidence would the benchmark be at the same elevation as the surface of the ground where the excavation is to be done. The actual depth of the excavation is the difference between the elevation at the surface of the ground and the elevation at the bottom of the excavation, **Figure 10–6.**

Concrete footings are placed in the bottom of the excavation to support the entire weight of the building,

**Figure 10–7.** These footings are placed on unexcavated earth to reduce the chance of the soil compacting under them. This means that the excavation contractor must measure the depth of the excavation accurately. The footings may be *stepped,* as in **Figure 10–8,** to accommodate a sloping site. This requires measuring the depth at each step of the footing. Information about the footing design is found on the foundation plan and building elevations. The layout of the foundation walls and their footings is shown on the foundation plan. The foundation walls are shown by two solid lines with dimensions to indicate their sizes, **Figure 10–9.** A dotted line on each side of the foundation indicates the concrete footing. The size of the footing may be omitted when the plan is developed for use in several locations.

The depth of the foundation, including its footing, is shown on the elevations. To simplify calculating excavation depths, many architects indicate the elevations as key points along the footings, **Figure 10–10.** A section view through all or part of the building may show a typical depth, but it is wise to check all the elevations for steps in the footing. The footings may be shown on the elevation as a double or a single dotted line. In masonry foundations, steps in footings are usually in increments of 8 inches to conform to standard concrete block sizes.

Use the following procedure to measure differences in elevation, such as the depth of an excavation:

Step 1.  Set the instrument up on a tripod, and level it in a convenient location.

**Note:** A laser level is often used to measure elevations and depths of excavations. Setting up a laser level is similar to setting up a builder's level. You should follow the manufacturer's instructions and obey the safety precautions provided for the use of the laser level.

Step 2.  Have a partner hold the target rod on a known elevation, such as a benchmark or ground of known elevation, while you focus the telescope and note the reading on the target rod where the cross hairs focus.

Step 3.  Take a similar reading with the target rod at the bottom of the excavation. The difference in the two readings is the depth of the excavation, **Figure 10–11.**

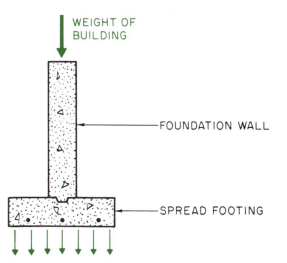

**Figure 10–7.** The footing spreads the weight of the building over a greater soil area.

**Figure 10–8.** Footings can be stopped to accommodate a sloping site.

FOUNDATION PLAN
SCALE ¼"=1'-0"

**Figure 10–9.** The foundation plan gives complete dimensions.

98.20
TOP OF MAS.

89.72

88.39

**Figure 10–10.** Spot elevations show key points on the footing and the foundation.

2'-4"

BENCH MARK OR
KNOWN ELEVATION

6'-10"

ON BOTTOM OF EXCAVATION

**Figure 10–11.** This excavation is 4'–6" deep—the difference between the two readings on the target rod.

## ✓ CHECK YOUR PROGRESS

Can you perform these tasks?

● Properly set up a builder's level.

● Use a builder's level and tape measure to lay out building lines.

● Check the accuracy of right angles using a tape measure only.

● Check the squareness of a building layout using a tape measure only.

● Calculate the depth excavation required for footings.

● Measure the depth of an excavation using a builder's level.

## ASSIGNMENT

Refer to the Lake House drawings (in the packet) to complete this assignment.

1. What is the distance from the Lake House to the nearest property boundary? (Do not treat the decks as part of the house for this question.)

2. What is the distance from the Lake House to the lake?

3. What is the distance from the north property line to the garage?

4. What is the area of the basement of the Lake House, including the foundation? Ignore slight irregularities in the shape of the foundation. For ease in calculating, divide the foundation into rectangles, **Figure 10–12.**

5. Find the highest and lowest natural grades meeting the house.

6. How much cut or fill is required for the basement at the northeast corner of the house?

7. Measuring from the natural grade, how deep is the excavation for the footing under the overhead garage door?

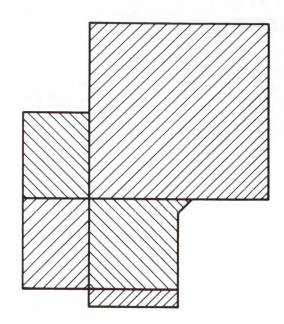

**Figure 10–12.** Use with assignment problem 4.

8. What is the elevation at the bottom of the deepest excavation for the Lake House? (Do not include the garage.)

9. Why would a row of large evergreen trees between the Lake House and the lake decrease the energy efficiency of the house?

10. What aspect of the location of a building is most often regulated by local ordinances?

# Site Utilities

## Objectives

After completing this unit, you will be able to perform the following tasks:

○ Interpret symbols and notes used to describe site utilities.

○ Explain the septic system indicated on a site plan.

○ Determine the pitch of drain lines.

## Sewer Drains

The *building sewer* carries the waste to the municipal sewer or septic system, **Figure 11–1.** Because sewer lines usually rely on gravity flow, they are large in diameter (4 inches, minimum) and are pitched to provide flow. Because water supply lines and gas lines are pressurized, pitch is not important in their installation. Therefore, the sewer is installed first, and other piping is routed around it as necessary. The size, material, and pitch of drains are usually given in a note on the site plan, **Figure 11–2.** The pitch of a pipe is given in fractions of an inch per foot. A pitch of ¼ inch per foot means that for every horizontal foot, the pipe rises or falls ¼ inch.

In some cases, sewers may have to flow uphill. This is the case with the Lake House. Uphill flow is accomplished by a *grinder pump,* **Figure 11–3.** A grinder pump grinds solids into small enough particles to be pumped and pumps the sewage at low pressure.

Figure 13-1 The house sewer carries waste from the house to

**Figure 11–1.** The house sewer carries waste from the house to the municipal sewer or septic systems.

4"C.I. @ ¼"

TO STREET
SEWER

DRIVE

**Figure 11–2.** This note indicates cast iron pipe sloped ¼ inch for every foot of run.

## Building Sewer

Any plumbing that is to be concealed by concrete work must be installed without fixtures or *roughed in* before the concrete is placed. Because the plumbing contractor installs all plumbing inside the building lines, this phase of construction is discussed later with mechanical systems. However, the sewer line from the building to the street is often installed before the building is erected, so it is discussed here.

The workers who install the sewer must be able to determine the elevation at which it passes through the foundation and the pitch of the line outside the building. The sewer line may be shown on plans as a solid or broken line. Although it is usually labeled, this is not always true. When the sewer is not labeled as such, it

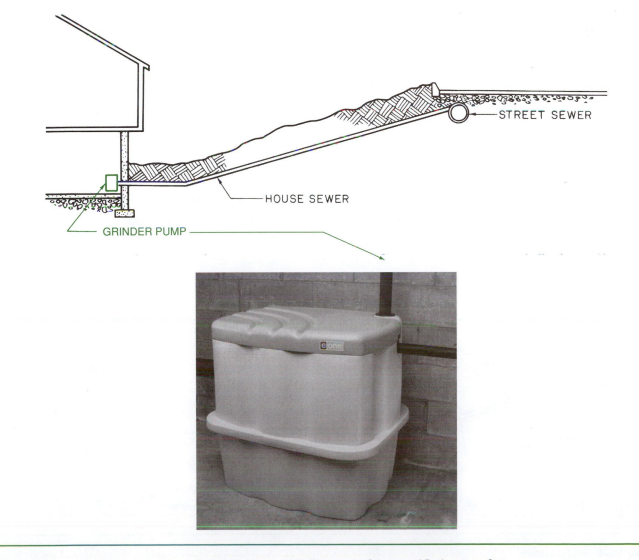

STREET SEWER

HOUSE SEWER

GRINDER PUMP

**Figure 11–3.** A grinder is a pump that can move sewage uphill. *Photo only: Courtesy of Environment One.*

can still be recognized by its material, pitch, and ending place. In light construction, the sewer is usually the only 4-inch pipe to the building. Also, the sewer is the only line with the pitch indicated.

## Municipal Sewers and Septic Systems

In most developed areas, the building sewer empties into a municipal sewer line near the street. The contractor for the new building is responsible for everything from the municipal sewer to the house.

In less developed areas, the sewer carries the sewage to a septic system. The most common type of *septic system* includes a septic tank and drain field, **Figure 11–4**. The septic tank holds the solid waste while it is decomposed by bacterial action. The liquids pass through the baffles and flow out of the tank to the distribution box. The distribution box (D.B.) diverts the liquid into *leach lines.* These perforated plastic or loose-fitting tile lines allow the liquid to be absorbed by the surrounding soil. The liquid gradually evaporates from or drains through the soil. The *drain field,* where the leach lines are laid, is usually made of a layer of crushed stone.

The design of septic systems is closely regulated by most health and plumbing codes. These local codes should be checked before designing or installing any

septic system. The building code or health department code often requires a *percolation test* before the system can be approved. In a *percolation test,* holes are dug. Then a measured amount of water is poured into each hole. The amount of time required for the water to drain into the soil is an indication of how well the soil will accept water from the septic system. This ability to accept water is called *percolation.* The locations of key elements in the system are often shown on the site plan.

## Other Piping

Other utility piping, such as that for water supply or gas, is shown on the site plan. If these lines will pass beneath the concrete footings or be concealed under a concrete slab, they must be roughed in before the concrete is placed. Water supply pipes follow the most direct route from the municipal water main or well to the main shutoff valve or pump. Gas lines run from the main to the gas meter. All supply lines on the plot plan should be labeled according to type and size of piping, **Figure 11–5.**

## Electrical Service

The electrical service is the wiring that brings electricity to the house. There are two types of electrical service:

**Figure 11–4.** Septic system.

DRAIN FIELD PER CITY HEALTH DEPT.

D.B.

1000 GAL. SEPTIC TANK

4" C.I. @ 1/4" PER FT.

FIN. FLOOR ELEV. 602.40

70'-0"

3/4" WATER

BURIED ELEC. SERVICE

50'-0"

3/4

18'-0"

G

HALLSTEAD ST.

**Figure 11–5.** Partial site plan with utilities indicated.

overhead and underground (or buried). Overhead service involves a cable from the utility company transformer or pole to a weather-tight fixture called a *service head* or *mast head* on the house, **Figure 11–6.** The service head is mounted on the top of a pipe, which serves as a conduit to the meter receptacle. In an underground service, the cable is buried, **Figure 11–7.** Although electrical service is a site utility, it is not usually installed until the building is enclosed.

The electrical service to residential and small commercial buildings is similar, as previously explained and as shown in **Figures 11–6** and **11–7.** This type of service has only one distribution point, the main electrical panel. Heavy commercial and industrial

SERVICE HEAD (BY OWNER)

OVERHEAD ELEC. SERVICE (BY UTILITY CO.)

POLE & TRANSFORMER (BY UTILITY CO.)

**Figure 11–6.** Overhead electrical service.

METER

BURIED CABLE

UP TO POLE
TRANSFORMER

OR

TO GROUND-LEVEL
TRANSFORMER

FROM UTILITY CO.

**Figure 11–7.** Underground electrical service.

electrical services require multiconductor feeders, and typically these are underground services from a utility transformer to a service entrance section in the designated electrical or utility room. Additional utility coordination will be required within the building, which is discussed in Unit 14. The placement of these underground electrical utility lines must be coordinated with the locations of other site and building utilities prior to starting installation.

## ✓ CHECK YOUR PROGRESS

Can you perform these tasks?

● Identify electrical service, water, and sewage lines shown on a site plan.

● Explain the operation of a septic system.

● Tell at what elevation sewer lines are to pass through a foundation.

● Tell what the elevation should be at each end of a pitched sewer pipe.

## ASSIGNMENT

Refer to the Lake House drawings (in the packet) to complete the assignment.

1. What size is the sewer for the Lake House?

2. How many lineal feet are required from the foundation wall to the septic tank?

3. What is the rise of the sewer from the house to the septic tank?

4. Where does the sewer pass through the foundation?

5. How many lineal feet of perforated pipe are needed for the drain field?

6. How many cubic yards of crushed stone are needed for the drain field?

7. What is the location of the electrical service?

# Footings

All soil can change shape under force. When the tremendous weight of a building is placed on soil, the soil tends to compress under the foundation walls and allow the building to settle. To prevent settling, concrete footings are used to spread the weight of the building over more area. The footings distribute the weight of the building, so that there is less force per square foot of area.

The simplest type of footing used in residential construction is referred to as *slab-on-grade.* In this system the main floor of the building is a single concrete slab, reinforced with steel to prevent cracking. This slab supports the weight of the building, **Figure 12–1.** Slab-on-grade foundations are common in warm climates. This type of construction is indicated on the floor plan by a note, **Figure 12–2,** and on section views of the construction, **Figure 12–3.** If excavation is involved in the construction where a slab-on-grade is to be placed, it is very important to thoroughly compact all loose fill before placing the concrete. Tamping the fill prevents the soil from compacting under the concrete later, causing the concrete to settle or crack.

A thickened slab, sometimes called a haunch, is used to further strengthen the slab where concentrated weight, such as a bearing wall, will be located. A *haunch* is an extra thick portion of the slab that is made by ditching the earth before the concrete is placed, **Figure 12–4.**

## Objectives

After completing this unit, you will be able to perform the following tasks:

- O Find all information on a set of drawings pertaining to footing design.

- O Interpret drawings for stepped footings used to accommodate changes in elevation.

- O Discuss applicable building codes pertaining to building design.

**Figure 12–1.** Slab-on-grade foundation.

4" CONC. SLAB w/6x6 10/10 WWF
OVER UNEXCAVATED
GRAVEL FILL w/6 MIL VAPOR BARRIER

12'-0"

2'-11"

3'-0½"

14'-0"

**Figure 12–2.** Note indicating slab-on-grade construction.

#235 ASPHALT SHINGLES
¾" O.S.B.
2-2x4
AL. DRIP EDGE
1 X 8 PINE
⅜" PLY. WD.
PLY. WD. SIDING
1" PLASTIC FOAM INSUL.
2x4 P.T.
6" MIN.
4" CONC. SLAB w/16"x12" HAUNCH. 2-#8 AT PERIMETER & 6x6-10/10 WWF IN SLAB

WD. TRUSS PER SPEC.
GYP. BD.
2x4 @ 16" O.C.
½" UNDERLAYMENT
6 MIL. VAPOR BARRIER
FIN. FLR.
4" CRUSHED STONE OR WASHED GRAVEL

TYPICAL WALL SECTION

SCALE: ⅜"=1'-0"

**Figure 12–3.** Typical wall section for slab-on-grade construction.

HAUNCH TO REINFORCE SLAB

CONCRETE SLAB

**Figure 12–4.** A haunch is a thickened part of a slab to reinforce it under a load-bearing wall.

## Spread Footings

In most sections of the country, the foundation of the house rests on a footing separate from the concrete floor. This separate concrete footing is called a *spread footing* because it spreads the force of the foundation wall over a wider area, **Figure 12–5.** Spread footings may be made by placing concrete inside wooden or metal forms, **Figure 12–6,** or placing the concrete in carefully measured ditches. In either case, the footing is shown on the foundation or basement plan by dotted lines outside the foundation wall lines, **Figure 12–7.**

FOUNDATION WALL

SPREAD FOOTING

**Figure 12–5.** A spread footing is so named because it "spreads" the downward force of the foundation wall over a greater soil area.

**Figure 12–6.** Footing forms.

8" CONC. FDN. (TYP.)

2'-0" X 10" CONC. FTG. (TYP.)

6"

3'-0"

2'-6"

9'-6"

5'-8"

**Figure 12–7.** The footing lines are sometimes shown on the foundation plan by broken lines around the foundation.

**Figure 12–8.** Anchoring the foundation wall to the footing.

The dimensions of the footings can be determined from the dimensions shown for the foundation. The foundation rests on the center of the footing unless otherwise specified, but the dimensions must meet the minimum requirements shown in Table R403.1 of the *International Residential Code®*. Therefore, if an 8-inch foundation rests on a 16-inch footing, the footing projects 4 inches beyond the foundation on each side. To lay out these footing lines, measure 4 inches from the building lines marked on the batter boards. Where footing lines cross to form a corner, suspend a plumb bob. Drive a stake under the plumb bob. Then, drive a nail in the stake to accurately mark the corner. When all corners are located in this manner, stretch a line between the nails to locate the inside of the footing forms.

A complete set of construction drawings also includes sections that show the spread footing in greater detail. However, these drawings are often superseded by local building codes. Building codes for footings include such things as minimum permissible depth of footing, required strength of concrete for footings, the width of footings, and the use of keys or dowels made of reinforcing steel, **Figure 12–8**. The drawings for the Lake House are drawn to satisfy the building codes that are enforced in the community where it is to be built. If the drawings were done for a plan catalog where the locality is not known in advance, the *International Residential Code®* might be used as a guide, but the builder would be referred to local building codes and many dimensions would be omitted. **Figure 12–9** shows an excerpt from the *International Residential Code®* section on footings.

When reading the foundation plan for footing dimensions, pay particular attention to special features like fireplaces and pilasters. (**Pilasters** are thickened sections of the foundation wall that add strength to the wall.) The footing in these areas will probably be wider and may be deeper than under straight sections of foundation wall. There are times when a footing or foundation wall must be crossed or penetrated by a utility line. Care must be taken when penetrating a footing or foundation wall. If the penetration is not shown on the drawings, prior approval by an engineer is most often required, and the building code probably regulates how it can be done. Items of concern include expansion and contraction between the utility line and footing or foundation wall, reduction in structural strength, and breaking the moisture barrier.

## Column Pads

Where steel columns, masonry piers, and wooden posts are used in the construction, a special concrete pad is indicated on the foundation plan, **Figure 12–10**. As with other footings, building codes for the area should be consulted for the design of these pads. These pads, as with all other footings, should rest on unexcavated or well-tamped earth.

## Reinforcement

Footings, column pads, and other structural concrete frequently include steel reinforcement. Footing reinforcement is normally in the form of steel reinforcement bars, commonly called *rebars*. Reinforcement bars are designated by their diameters in eighths of an inch, **Figure 12–11**.

**TABLE R403.1**
**MINIMUM WIDTH OF CONCRETE OR MASONRY FOOTINGS**
(inches)[a]

| | LOAD-BEARING VALUE OF SOIL (psf) | | | |
|---|---|---|---|---|
| | 1,500 | 2,000 | 3,000 | ≥ 4,000 |
| **Conventional light-frame construction** | | | | |
| 1-story | 12 | 12 | 12 | 12 |
| 2-story | 15 | 12 | 12 | 12 |
| 3-story | 23 | 17 | 12 | 12 |
| **4-inch brick veneer over light frame or 8-inch hollow concrete masonry** | | | | |
| 1-story | 12 | 12 | 12 | 12 |
| 2-story | 21 | 16 . | 12 | 12 |
| 3-story | 32 | 24 | 16 | 12 |
| **8-inch solid or fully grouted masonry** | | | | |
| 1-story | 16 | 12 | 12 | 12 |
| 2-story | 29 | 21 | 14 | 12 |
| 3-story | 42 | 32 | 21 | 16 |

For SI:   1 inch = 25.4 mm, 1 pound per square foot = 0.0479 kN/m².

a. Where minimum footing width is 12 inches, a single wythe of solid or fully grouted 12-inch nominal concrete masonry units is permitted to be used.

**Figure 12–9.** This is a table from the *International Residential Code®*. The entire section on footings is six pages long. *Reproduced with permission from the International Residential Code®, 2006. Falls Church, VA: International Code Council, 2003.*

4" PIPE COL.

24"x 24"x 10" CONC. FTG. (TYP.)

*PLAN*

PIPE COL.

CONCRETE FOOTING

*PICTORIAL*

**Figure 12–10.** Column footings appear as a rectangle of broken lines on the plan.

| REINFORCEMENT BARS | |
|---|---|
| Size Designation | Diameter in Inches |
| 3 | .375 |
| 4 | .500 |
| 5 | .625 |
| 6 | .750 |
| 7 | .875 |
| 8 | 1.000 |
| 9 | 1.128 |
| 10 | 1.270 |
| 11 | 1.410 |
| 14 | 1.693 |
| 18 | 2.257 |

**Figure 12–11.** Standard sizes of rebars.

# Depth of Footings

In many sections of North America, the moisture in the surface of the earth freezes in the winter. As this frost forms, it causes the earth to expand. The force of this expansion is so great that if the earth under the footing of a building is allowed to freeze, it either cracks the footing or moves the building. To eliminate this problem, the footing is always placed below the depth of any possible freezing. This depth is called the **frostline, Figure 12–12.**

The frost-depth map shown in **Figure 12–12** is only approximate and is not generally accurate enough for foundation design. **Figure 12–13** is from the *International Residential Code®* and is to be used by local building departments to specify foundation and footing design more precisely. The building department would fill in each of the spaces on the form for its particular jurisdiction.

Two methods are commonly used to indicate the elevation, or depth, of the bottom of the footings. The easiest to interpret is for spot elevations to be given at key points on the elevation drawings. Where elevations are given in this manner, the tops of the footing forms are leveled with a leveling instrument, using a benchmark for reference.

Another commonly used method is to dimension the bottom of the footing from a point of known elevation. This may be the finished floor or the top of the masonry

**Figure 12–12.** Average frost depths in the United States.

**TABLE R301.2(1)**
**CLIMATIC AND GEOGRAPHIC DESIGN CRITERIA**

| GROUND SNOW LOAD | WIND SPEED[e] (mph) | SEISMIC DESIGN CATEGORY[g] | SUBJECT TO DAMAGE FROM | | | | WINTER DESIGN TEMP[f] | ICE SHIELD UNDER-LAYMENT REQUIRED[i] | FLOOD HAZARDS[h] | AIR FREEZING INDEX[j] | MEAN ANNUAL TEMP[k] |
|---|---|---|---|---|---|---|---|---|---|---|---|
| | | | Weathering[a] | Frost line depth[b] | Termite[c] | Decay[d] | | | | | |
| | | | | | | | | | | | |

For SI: 1 pound per square foot = 0.0479 kN/m², 1 mile per hour = 1.609 km/h.

a. Weathering may require a higher strength concrete or grade of masonry than necessary to satisfy the structural requirements of this code. The weathering column shall be filled in with the weathering index (i.e., "negligible," "moderate" or "severe") for concrete as determined from the Weathering Probability Map [Figure R301.2(3)]. The grade of masonry units shall be determined from ASTM C 34, C 55, C 62, C 73, C 90, C 129, C 145, C 216 or C 652.

b. The frost line depth may require deeper footings than indicated in Figure R403.1(1). The jurisdiction shall fill in the frost line depth column with the minimum depth of footing below finish grade.

c. The jurisdiction shall fill in this part of the table with "very heavy," "moderate to heavy," "slight to moderate," or "none to slight" in accordance with Figure R301.2(6) depending on whether there has been a history of local damage.

d. The jurisdiction shall fill in this part of the table with "moderate to severe," "slight to moderate," or "none to slight" in accordance with Figure R301.2(7) depending on whether there has been a history of local damage.

e. The jurisdiction shall fill in this part of the table with the wind speed from the basic wind speed map [Figure R301.2(4)]. Wind exposure category shall be determined on a site-specific basis in accordance with Section R301.2.1.4.

f. The outdoor design dry-bulb temperature shall be selected from the columns of 97½-percent values for winter from Appendix D of the *International Plumbing Code.* Deviations from the Appendix D temperatures shall be permitted to reflect local climates or local weather experience as determined by the building official.

g. The jurisdiction shall fill in this part of the table with the Seismic Design Category determined from Section R301.2.2.1.

h. The jurisdiction shall fill in this part of the table with (a) the date of the jurisdiction's entry into the National Flood Insurance Program (date of adoption of the first code or ordinance for management of flood hazard areas), (b) the date(s) of the currently effective FIRM and FBFM, or other flood hazard map adopted by the community, as may be amended.

i. In accordance with Sections R905.2.7.1, R905.4.3, R905.5.3, R905.6.3, R905.7.3 and R905.8.3, for areas where the average daily temperature in January is 25°F (-4°C) or less, or where there has been a history of local damage from the effects of ice damming, the jurisdiction shall fill in this part of the table with "YES." Otherwise, the jurisdiction shall fill in this part of the table with "NO."

j. The jurisdiction shall fill in this part of the table with the 100-year return period air freezing index (BF-days) from Figure R403.3(2) or from the 100-year (99%) value on the National Climatic Data Center data table "Air Freezing Index- USA Method (Base 32° Fahrenheit)" at www.ncdc.noaa.gov/fpsf.html.

k. The jurisdiction shall fill in this part of the table with the mean annual temperature from the National Climatic Data Center data table "Air Freezing Index- USA Method (Base 32° Fahrenheit)" at www.ncdc.noaa.gov/fpsf.html.

**Figure 12–13.** Climate and geographic design information to be provided by the local building department in the *International Residential Code®. Reproduced with permission from the International Residential Code®, 2006. Country Club Hills, IL International Code Council, 2006.*

foundation, for example. These dimensions are given on the building elevations. Footings and other features marked with a reference symbol, called ⊕ a **datum** symbol, are to be used as reference points for other dimensions.

For an example, see the South Elevation 3/3 of the Lake House (included in the packet). The left end of this view shows a footing with its bottom at an elevation of 334.83 feet. This is a variation of the usual practice of showing the elevation of the top of the footing. The top of this footing is 10 inches higher, or 335.66 feet. (See Math Review 16.) What room of the Lake House is this footing under?

At the right end of the South Elevation 3/3, the footing is shown to be 5'-4" below the finished floor and masonry. This is the basement floor, which the Site Plan shows to be at 337.0 feet. Therefore, the top of this footing is at 331'-8", or 331.66'.

## Stepped Footings

On sloping building sites, it is necessary to change the depth of the footings to accommodate the slope. This is done by stepping the footings. When concrete blocks are to be used for the foundation walls, these steps are normally in increments of 8 inches. This allows the concrete blocks to be laid so that the top of each footing step is

**Figure 12–14.** Stepped footing.

even with a masonry course. Some buildings require several steps in the footing to accommodate steeply sloping sites. Steps in the footings are shown on the elevation drawings and on the foundation plan by a single line across the footing.

The Lake House has several steps in the footing. For example, see the east side of the garage. This is shown in the East Elevation 2/3. This step is also shown on the Foundation Plan. It is 8'-0" from the north end of the garage. Notice that the two levels of the footing overlap each other. These are built in one overlapping section, **Figure 12–14.**

---

## ✓ CHECK YOUR PROGRESS

Can you perform these tasks?

- List the dimensions of a slab-on-grade, including footings, when this type of foundation is shown on construction drawings.

- List the thickness, width, and reinforcement to be used for spread footings.

- Identify steps in footings, and give the dimensions of each step.

- Give the locations and dimensions of footings for columns, posts, and other features.

- Give the dimensions and reinforcement to be used for any thickened haunches in a concrete slab-on-grade.

## ASSIGNMENT

Refer to the Lake House drawings (included in the packet) to complete this assignment.

1. What is the typical width and depth of the concrete footings for the Lake House?

2. What is the total length and width (outside dimensions) of the concrete footings for the garage of the Lake House? (Remember to allow for the footings to project beyond the foundation wall.)

3. How many concrete pads are shown for footings under columns or piers in the Lake House?

4. What are the dimensions of these pads?

5. What reinforcement is indicated for these pads?

6. What reinforcement is indicated for the spread footings under the Lake House?

7. What is indicated by the 2-inch dimension between the 12-inch round concrete footings?

8. What is the elevation of the top of the footing under the garage door?

9. What are the elevations of the tops of each section of concrete footing shown on the East Elevation 2/3?

10. How far outside the foundation walls are the typical footings?

11. Refer to the building code in your community (or the model code section shown in **Figure 12–9**), and list the specific differences between the Lake House footings and the minimum code requirements. Assume a soil load-bearing capacity of 2,000 pounds per square foot.

# Foundation Walls

## Laying Out the Foundation

When the concrete for the footings has hardened and the forms are removed, carpenters can begin erecting forms for **concrete** foundations, or masons can begin laying blocks or bricks for masonry foundations. Although the material differs, the drawings and their interpretation for each type of foundation are similar.

In Unit 12 you referred to the dimensions on the foundation plan to lay out the footings. The same dimensions are used to lay out the foundation walls. The layout process is also similar. The outside surface of the foundation wall is laid out using previously constructed batter boards. Then the forms are erected or the masonry units are laid to these lines. The foundation plan includes overall dimensions, dimensions to interior corners and special constructions, and dimensions of special smaller features. It is customary to place the smallest dimensions closest to the drawing. The overall dimensions are placed around the outside of the drawing, **Figure 13–1.**

All drawing sets include, at least, a wall section showing how the foundation is built, how it is secured to the footings, and any special construction at the top of the foundation wall, **Figure 13–2.** Although a typical wall section may indicate the thickness of the foundation wall, you should carefully check around the entire wall on the foundation plan to find any notes that indicate varying thicknesses of the foundation wall. For example, the wall may be 12 inches thick where it has to support brick veneer above, while it is only 8 inches thick on the back of the building where there is no brick veneer. A careful check of the foundation plan for the Lake House shows that the house foundation calls for 12-inch, 10-inch, and 6-inch concrete block and that the garage foundation calls for 8-inch block.

Structural concrete is the most common foundation material, because it is strong and concrete foundations can be built quickly. The Lake House used for study in this unit uses concrete blocks for the foundation, because the drawings are similar for either material; however, there are some additional topics to be discussed for masonry (concrete block) foundations.

The details for a masonry foundation may call for horizontal reinforcement in every second or third course. This is usually prefabricated wire reinforcement to be embedded in the mortar joints. Prefabricated wire reinforcement is available in varying sizes for different sizes of concrete blocks.

The height of the foundation wall is dimensioned on the building elevations. These are the same dimensions as those used to determine the depth of the

## Objectives

After completing this unit, you will be able to perform the following tasks:

O Determine the locations and dimensions of foundation walls indicated on a set of drawings.

O Describe special features indicated for the foundation on a set of drawings.

**Figure 13–1.** Dimensioning on a foundation plan.

footings in the preceding unit. Just as the footing was stepped to accommodate a sloping building site, the top of the foundation wall may be stepped to accommodate varying floor levels in the *superstructure* (construction above the foundation), **Figure 13–3.**

The top of a masonry foundation may be built with smaller concrete blocks to form a ledge upon which later brickwork will be built. It is also common practice to use one course of 4-inch solid block as the top course of the foundation wall.

In concrete foundations, **anchor bolts** are placed in the top of the foundation, **Figure 13–4.** These bolts are left protruding out of the top of the foundation so that the wood superstructure can be fastened in place later. Anchor bolts are not normally shown on the foundation plan, but a note on the wall section indicates their spacing. On masonry walls, anchor bolts can be placed in the hollow cores of the concrete blocks. They are held in place

by filling the core with mortar grout. **Grout** is a **portland cement** mixture that has high strength.

In areas where there is a threat of extremely high winds or earthquakes, additional hold-down straps may be called for, **Figure 13–5.** These hold-downs are normally only used with concrete foundations.

## Special Features

Many foundations include steel or wooden beams, which act as girders to support the floor framing over long spans, **Figure 13–6.** When the girder is steel, it is indicated by a single line with a note specifying the size and type of structural steel. A wood girder is usually indicated by two or more lines and a note specifying the number of pieces of wood and their sizes in a built-up girder, **Figure 13–7.**

BASE & SHOE MOULD
FINISH FLOOR
BLDG. PAPER
SUB FLOOR

BLOCKING

FINISH OPTIONAL

8'-0" OR AS NOTED

2×4 STUDS 16" O.C.

TREATED PLATE
W/ 1/2" BOLTS 8'O.C.

GRADE

1" RIGID INSULATION

2 NO. 5 Ø RODS
WHERE REQUIRED

4" | 1'-0" | 4"
1'-8"

**Figure 13–2.** Section through foundation. *Courtesy of W. D. Farmer.*

**Figure 13–4.** Anchor bolt.

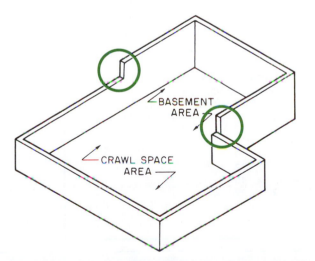

BASEMENT AREA

CRAWL SPACE AREA

**Figure 13–3.** The top of the foundation may be stepped to allow for a partial basement or varying floor levels.

**Figure 13–5.** Anchor bolts and hold-down strap are used in an earthquake zone.

**Figure 13–6.** The girder supports the floor framing. *Courtesy of the Trus Joist Corporation.*

COMMON SIZES OF STRUCTURAL STEEL GIRDERS

┌─ SHAPE:
│  S = STANDARD BEAM
│  W = WIDE-FLANGE BEAM
│  ┌─ NOMINAL DEPTH IN INCHES
│  │  ┌─ WEIGHT PER FOOT IN POUNDS

| S | 10 | × | 35 |
| S | 8 | × | 23 |
| S | 8 | × | 18.4 |
| W | 10 | × | 33 |
| W | 10 | × | 21 |
| W | 10 | × | 11.5 |
| W | 8 | × | 31 |
| W | 8 | × | 28 |
| W | 8 | × | 17 |
| W | 8 | × | 13 |

COMMON SIZES OF WOOD BUILT-UP GIRDERS

┌─ NUMBER OF PIECES
│  ┌─ THICKNESS EACH
│  │  (NOMINAL) IN INCHES
│  │  ┌─ WIDTH EACH (NOMINAL)
│  │  │  IN INCHES

| 2 | – | 2 | × | 10 |
| 2 | – | 2 | × | 12 |
| 3 | – | 2 | × | 10 |
| 3 | – | 2 | × | 12 |

**Figure 13–7.** Typical specifications for structural steel and wood built-up girders—other types are described in the project specifications.

AIR SPACE FOR EXPANSION

GIRDER

CONC. FOUNDATION

SECTION THRU BEAM POCKET

**Figure 13–8.** A beam pocket is a recess in the wall to hold the girder.

If the top of the girder is to be flush with the top of the foundation beam, pockets must be provided in the foundation, **Figure 13–8. Beam pockets** are usually shown on the details and sections of the construction drawings. The locations of these beam pockets are dimensioned on the foundation plans.

If windows are to be included in the foundation, the form carpenter or mason must provide rough openings of the proper size. The locations of windows should be dimensioned on the foundation plan. The sizes of the windows may be shown by a note or given on the window schedule. Window sizes and window schedules are discussed in Unit 25. It is important, however, to get the masonry opening size from the window manufacturer before forming the opening in the foundation wall. The *masonry opening* is the size of the opening required in the foundation wall to accommodate the window. This size may be different from the **nominal size** given in a note on the foundation plan. The foundation may include pilasters for extra support. A *pilaster* is a thickened section of the foundation, which helps it resist the pressure exerted by the earth on the outside. The location and size of pilasters are shown on the foundation plan.

The Lake House drawing includes a special feature not commonly found on foundation plans for houses. There are four notes that read 3½" ☐ STD. WT. STL. COL. W/8 × 8 × ½ B.PL. These notes indicate a 3½-inch square, standard-weight, steel column with 8-inch-by-8-inch-by-½-inch-thick base plates. This structural steelwork is explained in more detail later, but to completely understand the foundation plan, it is necessary to know that

the steel will be erected. The stress that this steelwork will place on the foundation is the reason for another note on the foundation plan in the area near the 18-inch round concrete footings. That note says FILL CORE SOLID W/ GROUT 1:1:6 = PORTLAND : MAS. CEMENT : AGGREGATE. The cores in the concrete block foundation where the steelwork will be erected are filled with this grout to provide the extra strength necessary to support the steel.

## Permanent Wood Foundation

Foundations are usually constructed of concrete or concrete block. However, a type of specially treated wood foundation is sometimes used, **Figure 13–9.** These *permanent wood foundations* do not use concrete footings.

**Figure 13–9.** Permanent wood foundation. *Courtesy of National Forest Products Association.*

2-2X6 P.T. PLATE

12"

PLYWD. STRIP HOLDS TOP OF POLYETHYLENE

6" FIBERGLASS BATTS

2X8 WALL STUD @ 16" O.C.

6 MIL POLYETHYLENE FILM

INT. FIN.

¾" P.T. PLYWOOD

6 MIL POLYETHYLENE FILM

½" EXP. JT.

2X6 P.T. BOTT. PLATE

2X8 P.T. FTG

6 MIL POLYETHYLENE FILM

12" GRAVEL FILL

**Figure 13–10.** Section of wood foundation.

Instead, they are built on 2 × 8s or laid on gravel fill below the frostline. The foundation walls are framed with lumber that has been pressure treated to make it rot resistant and insect resistant. The framing is covered with a plywood skin, and the plywood is covered with polyethylene film for complete moisture proofing, **Figure 13–10.** All the nails in a wood foundation are stainless steel to prevent rusting.

## ✓ CHECK YOUR PROGRESS

Can you perform these tasks?

● Give the length, width, and any offsets in a foundation wall.

● Name the material to be used for the foundation walls, including any reinforcement.

● Describe the locations and dimensions of all window openings, door openings, beam pockets, and other openings in the foundation walls.

● Tell what the elevation is at any point on the foundation wall.

● Describe anchor bolts and other tie-downs to be embedded in the foundation.

## ASSIGNMENT

Refer to the Lake House drawings (in the packet) to complete the assignment.

1. What is the typical thickness of the concrete block foundation for the Lake House?

2. How many lineal feet of block wall are included in the foundations of the Lake House and garage?

3. How thick is the south foundation of the fireplace?

4. What is the elevation of the top of the north end of the east foundation wall of the Lake House?

5. What is the highest elevation on the Lake House foundation?

6. How many courses of block are required at the highest elevation of the Lake House foundation?

7. What size anchor bolts are indicated at the top of the Lake House foundation?

8. What spacing is indicated for the anchor bolts?

9. How close (in courses) is horizontal reinforcement to be placed in the Lake House foundation?

10. What is the elevation of the top of the concrete block wall at the southwest 3½" ◻ steel column?

11. In how many places are the concrete block walls to be filled with grout?

# Drainage, Insulation, and Concrete Slabs

## Drainage

After the foundation walls are erected and before the excavation outside the walls is **backfilled** (filled with earth to the finished grade line), footing drains, if indicated, must be installed. **Footing drains** are usually perforated plastic pipe placed around the footings in a bed of crushed stone, **Figure 14–1.** If the site has a natural slope, the footing drains can be run around the foundation wall to the lowest point, then away from the building to drain by gravity. In areas where there is no natural drainage, the drain is run to a dry well or municipal storm drain.

At one time, clay drain tile was the most common type of pipe for this purpose. However, perforated plastic pipe is used in most new construction. Plastic drain pipe is manufactured in 10-foot lengths of rigid pipe and in 250-foot rolls of flexible pipe, **Figure 14–2.** An assortment of plastic fittings is available for joining rigid plastic pipe. When footing drains are to be included, they are shown on a wall section or footing detail, **Figure 14–3.** A note on the drawing indicates the size and material of the pipe.

If the floor drains are to be included in concrete-slab floors, they are indicated by a symbol on the appropriate floor plan, **Figure 14–4.** If these floor

## Objectives

After completing this unit, you will be able to perform the following tasks:

○ Locate and explain information for control of groundwater as shown on a set of drawings.

○ Locate and describe subsurface insulation.

○ Determine the dimensions of concrete slabs and the reinforcement to be used in concrete slabs.

FILTER FABRIC

CRUSHED STONE OR GRAVEL

PERFORATED PLASTIC PIPE PITCHED TOWARD STORM SEWER, DRY WELL, OR SUMP

**Figure 14–1.** Footing drain.

**Figure 14–4.** The floor drain is shown by a symbol. (Notice that the floor is pitched toward the drain.)

8"X16" PILASTER

UTILITY RM

□ FD

8'-0"

¼" PER 1'-0"

8'-8"

8"

**Figure 14–2.** Plastic drain pipe.

2X10 JOISTS @ 16" O.C.

2X6 P.T. SILL

6" ANCHOR BOLT

10" CONC. BLK.

4"

ASPHALT COATING

#6 DOWELS @ 6'-0"

½" EXP. JT.

4" CONC. FLR. W/6X6-10/10 WWF

12" GRAVEL FILL

4" PER. PLAS. PIPE @ ⅛" PER 1'-0"

2'-0"X10"CONC. FTG. W/2-#4s

5" GRAVEL FILL

POLYETHYLENE FILM

TYP. SECTION THRU FOUNDATION

SCALE: ½" = 1'-0"

**Figure 14–3.** Footing drains are shown outside the footing.

drains run under a concrete footing, the piping had to be installed before the footing was placed (see Unit 11). The **riser** (vertical part through the floor) and drain basin are usually set at the proper elevation just prior to placing the concrete floor.

The floor plan may include a spot elevation for the finished drain, or it may be necessary to calculate it from information given for the pitch of the concrete slab. Pitch of concrete slabs is discussed later in this unit.

## Vapor Barriers

Another technique often used to prevent groundwater from seeping through the foundation is coating the foundation wall with asphalt foundation coating.

At this point, subsurface work outside the foundation wall is completed, but backfilling should not be done until the superstructure is framed. The weight and rigidity of the floor on the foundation wall help the wall resist the pressure of the backfill. If the backfilling must be done before the framing, the foundation walls should be braced. To retard the flow of moisture from the earth through the concrete-slab floor, the drawings may call for a layer of gravel over the entire area before the concrete is placed. A polyethylene **vapor barrier** is laid over the gravel underfill. The thickness of polyethylene sheeting is measured in **mils**. One mil equals 1/1,000 of an inch. For vapor barriers, 6-mil polyethylene is generally used.

## Insulation

In cold climates, it is desirable to insulate the foundation and concrete slab. This insulation is usually rigid plastic foam board placed against the foundation wall or laid over the gravel underfill, **Figure 14–5.**

Like all materials, concrete and masonry expand and contract slightly with changes in temperature. To allow for this slight expansion and contraction, the joint between the concrete slab and foundation wall should include a compressible expansion joint material. Expansion joints can be made from any compressible material such as neoprene or composition sheathing material. This expansion joint filler is as wide as the slab is thick and is simply placed against the foundation wall before the concrete is placed.

## Concrete Slabs

When the house has a basement, the floor is a concrete slab-on-grade. The areas to be covered with concrete are indicated on the foundation plan or basement floor plan. This is usually done by an area not giving the thickness of the concrete slab and any reinforcing steel to be used. To help the concrete resist minor stresses, it is usually reinforced with welded wire fabric. The specifications for welded wire fabric are explained in **Figure 14–6.** Where

**Figure 14–5.** Rigid plastic foam insulation may be laid under the perimeter of the floor or against the foundation wall.

8 X 8 - 10/10 WWF

8"X 8"
SPACING ——

——WELDED WIRE FABRIC

——10 GA. WIRE BOTH WAYS

**Figure 14–6.** The callout for welded wire fabric explains the size and spacing of the wires.

the slab must support bearing walls or masonry partitions, it may be haunched, as discussed in Unit 12.

When floor drains are included or where water must be allowed to run off, the slab is *pitched* (sloped slightly). A note on the drawings indicates the amount of pitch. One-quarter inch per foot is common. When there is any possibility of confusion about which way the slab is to be pitched, bold arrows are drawn to show the direction the water will run, **Figure 14–7.**

When floor drains or forms are set for pitched floors, it is necessary to find the total pitch of the slab. This is done by multiplying the pitch per foot by the number of feet over which the slab is pitched. (See Math Review 8.) For example, if the note on a concrete **apron** in front of a garage door indicates a pitch of ½ inch per foot and the apron is 4 feet wide, the total pitch is 2 inches. The proper elevation for the form at the outer edge of the apron is 2 inches less than the finished floor elevation.

Slab-on-grade installations may require electrical raceways to be installed in or just below the concrete slab in the crushed or gravel fill. These installations must be made so that they do not reduce the structural integrity of the concrete slab. An oversized raceway in the concrete slab may cause the slab to crack and settle unevenly. An electrician should be present during concrete placement to observe and correct any damage to these electrical raceways.

All underground utility systems must be coordinated prior to starting installation, especially in commercial and industrial buildings, as discussed in Unit 11. The service entrance section may have up to six subdistribution sections or panels located throughout the building. Electrical feeders are required from the electrical service entrance section to these subdistribution sections or panels. This requires detailed coordination between the electrical installer and the other utility installers prior to starting the installation of the utility systems. The electrical installation must be laid out around the other utilities, with the other utility layouts normally having priority. The other utilities to be coordinated may include plumbing, fire sprinkler, heating and air conditioning, and specialty systems. This coordination must be done for both the underground and aboveground systems.

5" CONC. FLR.
PITCH 1/8" PER FOOT

O.H. GAR. DR.

O.H. GAR. DR.

**Figure 14–7.** A bold arrow indicates the way that water will run off a pitched surface.

Some of the concrete work in the Lake House is of particular interest, because it is part of the passive solar heating system the Lake House uses. Section 1/4 and the lower-level floor plan 1/2 indicate that the area under the living room and dining room floors is a heat sink. A *heat sink* is a mass of dense material that absorbs the energy of the sun during the day and radiates it at night. The living room and dining room are on the south side of the Lake House. In the winter, when the leaves are off the deciduous trees, the sun shines in the large areas of glass in these rooms and warms the heat sink. At night, this heat is radiated into the house to provide additional heat when it is needed most. The floor over the heat sink is a concrete slab similar to that used in the playroom. Detail 2/5 helps explain this concrete slab.

## ✓ CHECK YOUR PROGRESS

Can you perform these tasks?

- Describe the footing drains shown on a set of construction drawings.

- Name the material and its thickness when foundation insulation is shown.

- Describe any vapor barriers or foundation coating to be applied to the foundation or under concrete slabs.

- List the dimensions of any concrete slab floors shown on the drawings.

- Describe the reinforcement to be used in concrete slabs.

- Explain the pitch of a concrete slab to provide for drainage.

## ASSIGNMENT

Refer to the Lake House drawings (in the packet) to complete this assignment.

1. What is the thickness of the concrete slab over the heat sink in the Lake House? What is the largest raceway (conduit) that could be installed in this slab?

2. Describe the reinforcement used in the concrete slab in the playroom of the Lake House.

3. How many square feet of 2-inch rigid insulation are needed for the Lake House heat sink?

4. What prevents moisture from seeping through the concrete slab floor in the Lake House?

5. What is the finished floor elevation of the Lake House garage?

6. What is the elevation of the floor drain in the utility room of the Lake House?

7. What is the purpose of the 8-inch-thick concrete haunch in the middle of the Lake House slab?

8. How many cubic yards of concrete are required for the garage floor? The basement floor?

# Framing Systems

## Objectives

After completing this unit, you will be able to identify each of the following types of framing on construction drawings:

O Platform

O Balloon

O Post-and-beam

O Energy-saving

## Platform Framing

**Platform framing,** also called **western framing,** is the type of framing used in most houses built in the last 30 years, **Figure 15–1.** It is called platform framing because as the rough floor is built at each level, it forms a platform on which to work while erecting the next level, **Figure 15–2.**

A characteristic of platform framing is that all wall *studs,* the main framing members in walls, extend only the height of one story. Interior walls, called *partitions,* are the same as exterior walls. The bottoms of the studs are held in position by a *bottom* (or *sole*) plate. The tops of the studs are held in position by a *top plate.* Usually, a *double top plate* is overlapped at the corners to tie intersecting walls and partitions together, **Figure 15–3.** In some construction, the second top plate is not used. Instead, metal framing clips are used to tie

**Figure 15–1.** Platform, or western, framing.

**Figure 15–2.** Platform, or western, framing provides a convenient work surface during construction.

## Balloon Framing

In **balloon framing,** the exterior wall studs are continuous from the foundation to the top of the wall, **Figure 15–5.** Floor framing at intermediate levels is supported by *let-in ribbon boards.* This is a board that fits into a notch in each joist and forms a support for the joists. Although balloon framing is not as widely used as it once was, some balloon-framing techniques are still used for special framing situations.

In both platform-frame and balloon-frame construction, the structural frame of the walls is covered with sheathing. **Sheathing** encloses the structure and, if a structural grade is used, prevents wracking of the wall. *Wracking* is the tendency of all the studs to move, as in a parallelogram, allowing the wall to collapse to the side, **Figure 15–6.** There are two ways to prevent wracking. Plywood or other structural sheathing at the corners of the building prevents this movement. Also, diagonal braces can be attached to the wall framing at the corners to prevent wracking, **Figure 15–7.**

**Figure 15–3.** The double top plate overlaps at the corners.

intersecting walls together. Upper floors rest on the top plate of the walls beneath. The framing members of the upper floors or roof are positioned over the studs of the wall that supports them.

Platform construction can be recognized on wall sections, **Figure 15–4.** Notice that the studs extend only from one floor to the next.

6.5

6"MTL.
TRUSS (SEE DET.)

6.3

6" MTL. STUDS

R19 INSUL.

8'-0½"

5/8" PART. BD. UNDLYMT.

½" CDX

2ND FLR.

10" STL. C JOISTS @ 16" O.C.

6.2

11"

8'-11½"

5/8" PART. BD. UNDLYMT

½" CDX

1ST. FLR.

10" STL. C JOISTS @ 16" O.C.

6.1

11"

2X6

10" CONC. FDN.

8'-3½"

4" CONC.

5.2

BASEM'T

TYP. WALL SECTION
SCALE: ¼" = 1'-0"

**Figure 15–4.** This can be recognized as platform construction because the studs extend only from one floor to the next.

Figure 15-5. Balloon framing.

Figure 15-6. Wracking.

Seismic codes (building codes for earthquake protection) often call for interior shear walls. These are framed walls covered with structural sheathing, **Figure 15-8.** Interior shear walls make the building stronger.

## Post-and-Beam Framing

Platform framing and balloon framing are characterized by closely spaced, lightweight framing members, **Figure 15-9.** *Post-and-beam* framing uses heavier framing members spaced farther apart, **Figure 15-10.** These heavy timbers are joined or fastened with special

**Figure 15–7.** Let-in bracing prevents wracking.

**Figure 15–8.** Interior shear walls may be used in earthquake zones.

hardware, **Figure 15–11.** Because post-and-beam framing uses fewer pieces of material, it can be erected more quickly. Also, although the framing members are large (ranging from 3 inches by 6 inches to 5 inches by 8 inches), their wider spacing results in a savings of material. However, to span this wider spacing, floor and roof decking must be heavier. Post-and-beam framing is sometimes left exposed to create special architectural effects, **Figure 15–12.**

The structural core of the Lake House uses posts and beams, detail 6/6. However, this is not purely post-and-beam construction. The posts are 3½-inch square steel

7 1/4"

16"

16"

16"

**Figure 15–9.** Conventional framing. 16" O.C.

tubing, and the beams are plywood box beams. These are properly called **beams** because they carry a load without continuous support from below. These beams are supported only in the beam pockets on the steel posts, **Figure 15–13.** In the Lake House the post-and-beam construction does not include exterior walls. In pure post-and-beam construction, the exterior walls have widely spaced posts, and the space between is filled in with nonload-bearing curtain walls. These curtain walls are merely panels that fill in the space between the structural elements—the posts and beams.

In areas where hurricanes and high tides are a threat, some houses are built as pole structures. Pole buildings are a variation of post-and-beam construction. Poles, which are treated to be insect and rot resistant, are set several feet in the ground and 8 to 12 feet apart. A *band joist,* or *header,* is then bolted to these poles, **Figure 15–14.** The floors, walls, and roof are framed within the pole structure, **Figure 15–15.** Pole buildings are strong; they resist severe winds. Pole construction allows buildings to be kept above damaging floodwaters.

## Metal Framing

Light-gauge metal framing is used extensively in commercial construction where the strength of structural steel or reinforced concrete is not required. Metal framing is also very popular in home building where termites and water

CEILING JOISTS ARE
OFTEN OMITTED FOR
SLOPING CEILING

**Figure 15–10.** Post-and-beam framing. 4' O.C.

damage are known to be problems. Light-gauge steel framing uses studs and joists, the same as wood framing. Unlike wood framing, steel framing uses tracks and runners in place of bottom plates and top plates, **Figure 15–16.** The lighter, less expensive grades of metal framing are not as strong as wood, so they are used primarily for interior partitions or in conjunction with other structural materials. Metal framing can be pre-assembled just as you would pre-assemble wood framing. The runner is fastened to the floor with powder-actuated fasteners. The wallboard material is fastened to the metal studs using an electric screw gun with sheet metal screws. Subflooring material is fastened to the metal floor joists similarly with power-driven screws. The construction drawings and specifications will indicate the type of metal framing to be used. The gauge must be taken into consideration before using certain tools (stud punches, powder-actuated tools, etc.) and special mounting devices (box hangers, conduit brackets, etc.).

Where electrical nonmetallic-sheathed cable is installed through metal framing, a bushing or grommet must be installed prior to installing the cable, **Figure 15–17.**

## Energy-Saving Techniques

When wall framing is done with 2 × 4s spaced 16 inches on centers, up to 25 percent of the wall is solid wood. Wood conducts heat out of the building. Only the space between the solid wood framing can be filled with insulation. By using 2 × 6 studs spaced 24 inches on centers, the area of solid wood is reduced to less than 20 percent of the wall. The amount of framing material is the same.

**Figure 15–11.** Common timber fastenings.

Not only does this reduce the amount of wood exposed to the surface of the wall, but it also allows for 2 inches more insulation.

The area of exposed wood is further reduced by special corner construction. In conventional framing, three pieces are used to frame the corner of a wall, **Figure 15–18**. To reduce heat loss through the wall, only two pieces are used, **Figure 15–19**. The third piece, which normally provides a nailing surface for the interior **drywall,** is replaced by metal clips.

Examination of the first floor plan and the detail drawings of the Lake House shows that the walls are framed for maximum efficiency. Several details indicate that the studs are 2 × 6s @ 24 O.C. This allows room for more insulation in the wall. Also notice that the house is sheathed with 3/4-inch insulation sheathing. Interior partitions do not need insulation, and so they are framed with 2 × 4s.

When installing nonmetallic-sheathed cable through wood wall studs and floor joists, the cable must be installed at least 11/4 inches from the nearest edge. Where this 11/4 inches cannot be met, a 1/16-inch-thick metal plate must be installed to protect the cable from nails and screws, **Figure 15–20**.

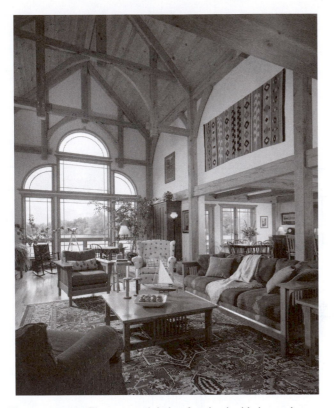

**Figure 15–12.** The exposed timber framing in this house is an important part of its design. *Courtesy of Riverbend Timber Framing, Inc. Photo by Roger Wade.*

WELDED

WELDED

WELDED

BOX BEAM

**Figure 15–13.** Beam pocket for the Lake House.

JOIST HEADER

POLE NOTCHED OR FACED

PRESSURE TREATED POLES

BAND JOIST OR HEADER

DRILLED HOLE

EMBEDMENT AND ALIGNMENT OF POLES. DEPTH OF EMBEDMENT DEPENDS ON SPACING AND SIZE OF POLES, WIND LOADS, AND SO FORTH, AND MAY VARY FROM 5 TO 8 FEET.

**Figure 15–14.** Basic elements of a pole building. *Courtesy of American Wood Preservers Institute.*

**Figure 15–15.** Framing in a pole building. *Courtesy of American Wood Preservers Institute.*

**Figure 15–16.** Metal framing. *Courtesy of Dietrich Metal Framing.*

**Figure 15–17.** Nonmetallic-sheathed cable installed through metal framing.

**Figure 15–18.** Conventional corner posts for 2 × 4 framing.

**Figure 15–19.** Corner construction for Arkansas Energy Saving System.

DRYWALL CLIP

1¹/4"

Less than 1¹/4"

Metal plate ¹/16" thick

2 x 4

**Figure 15–20.** Nonmetallic-sheathed cable installed in wood framing.

## ✓ CHECK YOUR PROGRESS

Can you perform these tasks?

● Identify platform framing shown on drawings, and describe how it differs from other types of framing.

● Identify balloon framing shown on drawings, and describe how it differs from other types of framing.

● Identify post-and-beam framing shown on drawings, and describe how it differs from other types.

● Identify the following framing members on a wall or building section: joists, bottom or sole plate, top plate, stud, ribbon (to support joists), posts, beams, and sheathing.

● Explain what prevents a building frame from wracking.

● Describe at least three techniques for reducing the amount of heat lost through solid wood in a building frame.

## ASSIGNMENT

1. Identify *a* through *g* in **Figure 15–21**.

2. What kind of framing is shown in **Figure 15–21**?

3. Identify *a* through *c* in **Figure 15–22**.

4. Identify *a* through *c* in **Figure 15–23**.

5. Sketch a plan view of a conventional corner detail. Include drywall and sheathing.

6. Sketch a plan view of an energy-efficient corner detail. Include drywall and sheathing.

7. What two materials are most often used for framing homes?

8. What framing material requires bushings or grommets for nonmetallic-sheathed cable installations?

Refer to the Lake House drawings (in the packet) to complete the rest of the assignment.

9. Which of the types of framing discussed in this topic is used for the Lake House?

10. What supports the west ends of the floor joists in bedroom #1?

11. What supports the east ends of the kitchen rafters?

12. How are the box beams fastened to the 3½-inch square posts?

13. How are the steel posts anchored?

14. What does the northwest square steel post rest on?

15. What supports the north edge of the living room floor?

16. List the dimensions (thickness × depth × length) of all box beams.

(**Note:** The length can be found by subtracting the outside dimension of the posts from the centerline spacing shown on the plan views.)

17. List the length of each piece of 3½-inch steel.

Figure 15–21.

Figure 15–22.

Figure 15–23.

# Columns, Piers, and Girders

The most common system of floor framing in light construction involves the use of joists and girders. **Joists** are parallel beams used in the floor framing, **Figure 16–1**. Usually buildings are too wide for continuous joists to span the full width. In this case, the joists are supported by one or more **girders** (beams) running the length of the building. The girder is supported at regular intervals by wood or metal posts or by masonry or concrete piers.

## Columns and Piers

Metal posts called *pipe columns* are the most common supports for girders. However, masonry or concrete piers may be specified. The locations of columns, posts, or piers are given by dimensions to their centerlines. When metal or wooden posts are indicated, the only description may be a note on the foundation plan, **Figure 16–2**. This note may give the size and material of the posts only, or it may also specify the kind of bearing plates to be used at the top and bottom of the post, **Figure 16–3**. A *bearing plate* is a steel plate that provides a flat surface at the top or bottom of the column or post. If the girder is supported by masonry or concrete piers, a special detail may be included to give dimensions and reinforcement details.

**Figure 16–1.** Joist-and-girder floor framing. *Courtesy of Richard T. Kreh, Sr.*

## Objectives

After completing this unit, you will be able to perform the following tasks:

O Locate columns and piers, and describe each from drawings.

O Locate and describe the girders that support floor framing.

O Determine the lengths of columns and the heights of piers.

FOUNDATION PLAN
SCALE: 1/8"=1'-0"

**Figure 16–2.** Foundation plan with note for column and footing.

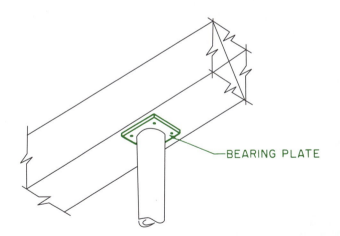

BEARING PLATE

**Figure 16–3.** Bearing plate.

## Girders

Before the total length of the columns or height of the piers can be calculated, it is necessary to determine the size of the girder and its relationship to the floor joist. Three types of girders are commonly used in residential construction. Steel beams are often used where strength is a critical factor. Built-up wood girders are constructed on the site. These consist of three or more pieces of 2-inch lumber nailed together with staggered joints to form larger beams. The sizes and specifications for built-up wood girders and structural steel girders are given in Table R502.5(1) of the *International Residential Code®* and shown on the foundation plan. Laminated veneer lumber (LVL), **Figure 16–4,** is also popular for use as girders. LVL is an engineered lumber product made by gluing layers of **veneer** together. These beams can be manufactured in almost any size. They are very strong, because lumber defects can be eliminated and all the grain runs in the best possible direction. The notes commonly found on a foundation plan to indicate the type of girder to be used are shown in **Figure 16–1.**

## Determining the Heights of Columns and Piers

The length of the columns or height of the piers depends on how the joists will be attached to the girder. The floor joists may rest directly on top of the girder or may

**Figure 16–4.** LVL (laminated veneer lumber) beams are popular, because they are strong and can be manufactured in any length and depth. *Courtesy of Trus Joist Macmillan, A limited Partnership, Boise, ID.*

be butted against the girder so that the top surface of the floor joist is flush with the top surface of the girder, **Figure 16–5.**

To find the height of the columns or piers, first determine the dimension from the basement floor to the finished first floor. Then subtract from this dimension the depth of the first floor, including all of the framing and the girder. Then add the distance from the top of the basement floor to the bottom of the column. The result equals the height of the column or pier. (See Math Reviews 5 and 6.) For example, the following shows the calculation of the height of the steel column in **Figure 16–6:**

○ Dimension from finished basement floor to finished first floor = 8′ – 10½″
Allowance for finished floor = 1″
Nominal 2 × 8 joists = 7¼″
2 × 4 bearing surface on girder = 1½″
W8 × 31 = 8″
Total floor framing = 17¾″ or 1′ – 5¾″

○ Subtract total floor framing
8′ – 10½″ minus 1′ – 5¾″ = 7′ – 4¾″

○ Add thickness of concrete slab
7′ – 4¾″ plus 4″ = 7′ – 8¾″

Figure 16–5. Several methods of attaching joists to girders.

Figure 16–6. Calculate the height of the steel column.

# ✓ CHECK YOUR PROGRESS

Can you perform these tasks?

● Give the dimensions to locate columns or piers to support floor framing.

● Describe columns, including their material, diameter, and bearing plates.

● Calculate the length or height of columns and piers.

● Describe the girders to be used in a building, including their material, depth, weight or thickness, and length.

● Explain how the floor joists bear on the girders.

## ASSIGNMENT

Questions 1 through 5 refer to **Figure 16–7.**

1. What is the length of the girder?

2. Describe the material used to build the girder, including the size of the material.

3. What supports the girder? (Include material and cross-sectional size.)

4. How many posts, columns, or piers support the girder?

5. What is the height of the columns or piers supporting the girder, including bearing plates?

Questions 6 through 10 refer to **Figure 16–8.**

6. What is the length of the girder?

7. Describe the material used to build the girder, including the size of the material.

8. What supports the girder? (Include material and cross-sectional size.)

9. How many posts, columns, or piers support the girder?

10. What is the dimension from the top of the footing under the pier to the top of the steel beam?

½" PLYWD.

2X4 @16"O.C.

½" CDX PLYWD

2X4

2X10

2X6 P.T.

2X6

4"X16"CONC. BLK.

3-2X10 P.T.

NOTCH
w/1" SPACE

5"

8"X16" CONC. BLK.

1'-4"X8" FTG.
W/2 - #6

A SECTION THRU BEAM POCKET
SCALE: ½"=1'-0"

2X10 @16"O.C.

9¾"

2X6

3-2X10

4"X4"X¼"
BEARING PL.
WELDED TOP
& BOTTOM

4'-0"

VAPOR
BARRIER

4"COL.

2'-0"X2'-0"X10"FTG.

B SECTION AT COLUMN
SCALE: ½"=1'-0"

56'-0"

8"

H.B.

UNEXCAVATED CRAWL SPACE
W/POLYETHYLENE MOISTURE BARRIER

4"CORBEL

12'-0"

4"CONC. SLAB
W/6X6 10/10 WWF

10'-8"     10'-8"     10'-8"     10'-8"

A

24'-0"

3-2X10

B

8"        8"

JOISTS ABOVE
2X10@16"O.C.

2'-8"X1'-4"MTL.
LOUVRE (TYP.)

12'-0"

4" PIPE COL.
W/4"X4"X¼"BEARING PLATE
2'-0"X2'-0"X10"CONC. FTG.
(TYP.)

12"

16"X8"CONC. FTG. (TYP.)

8"

12'-0"        4'-0"     8"     20'-0"     8"        3'-4"      22'-8"

H.B.

12'-0"                    44'-0"                    8"

FOUNDATION PLAN
SCALE: ⅛"=1'-0"

**Figure 16–7.** Use with assignment questions 1 through 5.

FOUNDATION PLAN
SCALE: 1/8"=1'-0"

52'-0"

6'-0"  3'-4"  8'-0"  3'-4"  8'-0"  3'-4"  4'-0"  16'-0"

2X10 JOISTS @2'-0"O.C.

10"    10"

12'-0"    12'-0"    12'-0"

W10X45

1'-8"X1'-8" CONC. BLK. PIER (TYP.)

3'-0"

2X10 JOISTS @2'-0"O.C.

4"CONC.FLR.

UP

16'-0"

36'-0"

10"

10"    10"

5"CONC. SLAB OVER UNEX. FILL

10"

20'-0"

8'-0"

Section at Pier & Foundation Wall
Scale 1/8" = 1'-0"

W10 x 45
2" x 6" P.T.

2" x 8" @ 16" O.C.
5/8" PLY. WD.

1'-7 5/8"

7'-1 1/2"

4"

1" Air Space
4" Solid Conc. Blk
2" x 6" P.T.
10" x 16" Conc. Blk.

2" x 6" P.T.
8" x 16" Conc. Blk.
Top Course Solid

Polyethylene
4" Conc.

1/2" EXP. JT.
TYP

2'-6" x 2'-6" x 8" Ftg.
W/3 – #4 Ea. Way

1'-14" x 8" Ftg.
W/2-#4

Note: Nominal Size to Actual Size Lumber

| Nominal | Actual |
|---------|--------|
| 2 x 4 | 1 1/2 x 3 1/2 |
| 2 x 6 | 1 1/2 x 5 1/2 |
| 2 x 8 | 1 1/2 x 7 1/4 |
| 2 x 10 | 1 1/2 x 9 1/4 |
| 2 x 12 | 1 1/2 x 11 1/4 |
| 2 x 14 | 1 1/2 x 13 1/4 |

**Figure 16–8.** Use with assignment questions 6 through 10.

# UNIT 17

# Floor Framing

## Objectives

After completing this unit, you will be able to perform the following tasks:

O Describe the sill construction shown on a set of drawings.

O Identify the size, direction, and spacing of floor joists according to a set of drawings.

O Describe the floor framing around openings in a floor.

## Sill Construction

Where the framing rests on concrete or masonry foundation walls, the piece in contact with the foundation is called the *sill plate,* **Figure 17–1.** The sill plate is the piece through which the anchor bolts pass to secure the floor in place. To prevent the sill plate from coming in direct contact with the foundation, and to seal any small gaps, a **sill sealer** is often included. This is a compressible, fibrous material that acts like a gasket in the sill construction. All lumber in contact with the foundation, including the sill plate, must be pressure treated (P.T.) with an approved chemical to prevent decay.

The entire construction of the floor frame at the top of the foundation is called *sill construction* or the **box sill.** The box sill is made up of the sill sealer, sill plate, joist, and joist header, **Figure 17–2.** The sizes of materials are given on a wall section or sill detail. For areas where termites are present, a termite shield is included in the sill construction, **Figure 17–3.** A **termite shield** is a continuous metal shield that prevents termites from getting to the wood superstructure.

**Figure 17–1.** Sill plate.

Figure 17–2. Box sill.

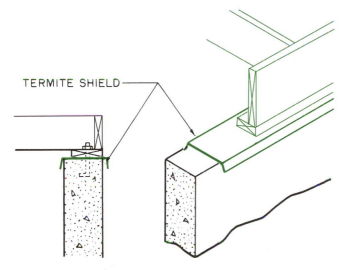

Figure 17–3. Termite shield.

# Floor Joists

*Floor joists* are the parallel framing members that make up most of the floor framing. Until recently, joists in residential construction were 2-inch framing lumber. However, recent advances in the use of materials have produced several types of engineered joists, **Figure 17–4.** Although the materials in each type are different, their use is essentially the same.

Metal floor joists are often used in commercial and industrial construction. Where nonmetallic-sheathed cable is installed through metal studs or floor joists, a bushing or grommet must be installed prior to installing the cable (see Figure 15–17).

Notes on the floor plans indicate the size, direction, and spacing of the joists in the floor above, **Figure 17–5.** For example, notes on the foundation plan give the information for the first floor framing. When the arrangement of framing members is complicated, a framing plan may be included, **Figure 17–6.** On most framing plans, each member is represented by a single line.

In the simplest building, all joists run in the same direction and are supported between the foundation walls by a single girder. However, irregularities in building shapes require that joists run in different directions, **Figure 17–7.** As the building design becomes more complex, more variations in floor framing are necessary. In a building such as the Lake House, which has floors at varying levels, the joists are supported by a combination of girders, beams, and load-bearing walls.

Figure 17–4. Several styles of joists.

**Figure 17–5.** Joist callouts for the first floor are shown on the foundation plan.

## Lake House Floor Framing

The floor framing for the Lake House is shown on Framing Plan 1/6 (included in the packet). Notice that the framing plan is made up of a simplified floor plan and single lines to represent floor joists, beams, and joist headers. A more elaborate type of framing plan uses double lines to represent the thickness of each member, **Figure 17–8**. The framing plan shows the location and direction of framing members, but for more detail it will be necessary to refer to Floor Plan 2/2 also.

The floor framing in the Lake House can be studied most easily if it is viewed as having four parts: the kitchen, bedroom #1, bedroom #2 and bathrooms, and the loft. There is no floor framing for the living room and dining room. These two areas form the heat sink for the passive-solar features discussed in earlier units. The floors in these rooms are concrete to absorb and radiate solar heat. As each area is framed, the carpenter must identify the following:

○ Joist headers (locate the outer ends of joists)

○ Bearing for inner ends of joists (beams, walls, etc.)

○ Size and type of framing materials

○ Length of joists

○ Spacing

○ Framing at openings

The joist headers are easily identified on the framing plan, but their exact position should be checked on the detail drawings. They may be set back to create a brick ledge, to accommodate wall finish, or to be flush with the foundation, **Figure 17–9**.

The *bearing* (support) for the inner ends of some of the joists in the Lake House is the structural-steel-and-box-beam core, detail 6/6. This core consists of four 3½″ square steel posts with wood box beams and steel-channel beams. The positions of the posts are shown on all plan views of the house. The C8 × 11.5 steel acts as a beam to support the floor in bedroom #2. Two 2 × 10s support the loft. The plywood box beams form part of the roof framing. The double 2 × 10 beam rests in beam pockets that are welded to the posts. The structural steel channel (C8 × 11.5) is bolted directly to the posts.

The inner ends of the remaining floor joists in the Lake House are supported by bearing walls in the lower

FLOOR FRAMING PLAN
SCALE: 3/16" = 1'- 0"

Within the plan, the following labels appear:

2X10 JOISTS @ 2'-0" O.C.

DOUBLE JOISTS & HDRS. AROUND OPENING

W10 X 39

2X10 JOISTS @ 2'-0" O.C.

W10 X 39

2X10 JOISTS @ 2'-0" O.C.

**Figure 17–6.** A simple floor framing plan.

level. For example, some of the kitchen floor joists are supported by the west wall of the playroom, near the fireplace; and some, by the 2 × 10 header that spans the distance from the playroom wall to the foundation in the crawl space.

The size and spacing of material to be used are given on the framing plan by a note. They can also be found on the wall section. Lengths of framing members are usually not included on framing plans. However, these lengths can be found easily by referring to the floor plan that shows the location of the walls or beams on which the joists rest. For example, refer to **Figures 17–10** and **17–11** and find the length of the floor joists in bedroom #2 as follows:

1. The dimension from the outside of the north wall to the centerline of the 3½" [icon] post is 14' – 1".

2. The dimension from the centerline of the 3½" [icon] post is 2", so the overall dimension of the bedroom floor is 14' – 3".

3. According to Wall Section ¾, the joist header is flush with the north foundation wall, so subtract 1½" (the thickness of the joist header) from each end.

4. 14' – 3" minus 3" (1½" at each end) equals 14' – 0" (the length of the joists).

Some floor framing is cantilevered to create a seemingly unsupported deck. *Cantilevered* framing consists of joists that project beyond the bearing surface to create a wide overhang. This technique is used extensively for balconies, under bay windows, and for garrison-style houses, **Figure 17–12.**

Floor Framing  117

**Figure 17–7.** Floor framing plan for an irregular-shaped house.

## Framing at Openings

Where stairs and chimneys pass through the floor frame, some of the joists must be cut out to form an opening. The ends of these joists are supported by headers made of two or more members. The full joists at the sides of the opening have to carry the extra load of the shortened joists and headers, so they are also doubled or tripled, **Figure 17–13.** The number of joists and headers required around openings of various sizes is spelled out in building codes and shown on framing plans.

**Figure 17–8.** Double-line framing plan.

½" SHEATHING

SIDING

8" BLOCK

GRADE

WATERPROOFING

4"  8"  4"
1'-4"

(A) FLUSH

½" SHEATHING

15 LB. FELT

METAL LATH & STUCCO

8" BLOCK

WATERPROOFING

4"  8"  4"
1'-4"

(B) STUCCO

BATT INSULATION

FOR JOIST SIZE SEE FOUNDATION PLAN

26 GA. G.I. TERMITE SHIELD OR APPROVED SOIL TREATMENT

INTEGRAL 4"X16" PILASTER NOT OVER 8'0.C.

4" CONC. SLAB W/NO. 10, 6"X6" W.W.M. AS REQ.

MEMBRANE OVER 4" GRAVEL

2 NO. 5 ∅ RODS

BRICK VENEER W/METAL TIES 16" VERT. & 32" HORIZ.

1" AIR SPACE

½" SHEATHING SEE SPEC.

FLASHING

WEEP HOLES

4" SOLID MASONRY

8"X8"X16" CONC. BLOCK

GRADE

WATERPROOFING OVER 3/8" PARGING

4" DRAIN TILE IN 6" GRAVEL

6"

2"

4"  16"
1'-8"

(C) BRICK VENEER

**Figure 17–9.** The joist headers are positioned according to the exterior finish to be used. *Courtesy of W. D. Farmer.*

**Figure 17–10.** Part of the Lake House Floor plan. *Courtesy of Robert C. Kurzon.*

**#3 SILL FLUSH W/FDN**

**Figure 17–11.** Part of the Lake House wall section. *Courtesy of Robert C. Kurzon.*

**Figure 17–12.** Typical uses of cantilevered framing.

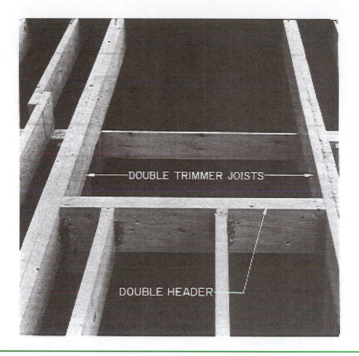

DOUBLE TRIMMER JOISTS

DOUBLE HEADER

**Figure 17–13.** Framing around an opening in the floor.

## ✓ CHECK YOUR PROGRESS

Can you perform these tasks?

● Describe the box sill shown on a set of drawings, including the size of the materials, anchoring, and setback from the face of the foundation.

● List the sizes of all the floor joists in a building.

● Describe the direction and spacing of all the floor joists in a building.

● Describe the framing around openings in the floor.

## ASSIGNMENT

Refer to the Lake House drawings to complete this assignment.

1. What size lumber is used for the floor framing in bedroom #2?

2. What size lumber is used for the floor joists in the loft?

3. What size lumber is used for the joist headers in the loft?

4. How long are the joists in the loft?

5. How long are the floor joists in bedroom #1?

6. What supports the west end of the floor joists in bedroom #1?

7. What supports the south end of the floor joists in bedroom #2?

8. How long are the floor joists in bedroom #2?

9. How many floor joists are needed for bedroom #2 and the adjacent closets?

10. What size material is used for the sill?

11. What does the double 2 × 10 beam in the structural core of the Lake House support?

12. How long are the headers that support the loft floor joists?

13. Is the box sill of the Lake House flush with the foundation wall or set back?

14. When an opening in a floor is framed, why are the joists at the sides of the opening doubled?

# UNIT 18

# Laying Out Walls and Partitions

## Objectives

After completing this unit, you will be able to perform the following tasks:

- Describe the layout of a house from its floor plans.

- Find specific dimensions given on floor plans.

When the deck (framing and subfloor or concrete slab) is completed, the framing carpenter lays out the location of walls and partitions. The size and location of each wall are indicated on the floor plans. Drawings 1/2 and 2/2 are the floor plans for the Lake House.

Of all the sheets in a set of construction drawings, the floor plans often contain the most information. Before looking for specific details on floor plans, it may help to mentally walk through the house. Start at the main entrance to the main floor, and visualize each room as if you were walking through the house.

## Visualizing the Layout of Walls and Partitions

The lowest floor with frame walls in the Lake House is the basement. Start at the $6^0 \times 6^8$ SGD (sliding glass door) on the east wall of the playroom. This large L-shaped room covers most of the basement floor. Some plans list overall dimensions for each room. The plans for the Lake House give this information by conventional dimension lines only.

The north-south dimension of the playroom is 28' – 3". The east-west dimension of the north part of the playroom is 14' – 1". The section of the playroom with the fireplace is 13' – 8" by 13' – 10". It may be helpful to notice the overall dimensions of each room as you visualize its shape and relationship to other rooms. It will be necessary to refer to these dimensions many more times. At the north end of the playroom is a small closet. South of the closet is a hall leading to the bathroom and utility room.

The bathroom has a shower, water closet (toilet), and lavatory. Notice that although most interior partitions are 3½ inches thick, the wall behind the water closet and lavatory is 5½ inches thick. This thicker wall is called a *plumbing wall.* Its extra thickness allows room for plumbing, **Figure 18–1.**

The purpose of the utility room beyond the bathroom is to house mechanical equipment such as the water heater, furnace or boiler, and water pump. In the southwest corner of the utility room is a small opening into the crawl space on the west side of the house. Notice that the concrete ends here.

Walking back through the playroom to the stairs, you will see some interesting features, **Figure 18–2.** A broken line in this area indicates the edge of a floor above. These floors can be seen more easily on the first floor plan. Also notice the location of the fireplace, which will extend up through the upper floor. The L-shaped stairs lead up to the first floor. It is obvious that this is not a full story higher because only four steps are shown. This can be seen more

Figure 18–1. Six-inch plumbing wall.

2X6 STUDS

4" PIPING

Figure 18–2. Lake House playroom.

Figure 18–3. Lake House living room.

Figure 18–4. Lake House kitchen.

Figure 18–5. Lake House loft.

clearly in the sections on Sheet 4. The L-shaped stairs lead up to the living room, **Figure 18–3.** Against the west and south walls is a plywood platform or built-in bench.

Another set of stairs in the northwest corner of the living room leads to the kitchen and dining room. These rooms are separated only by a peninsula of kitchen cabinets. The kitchen is separated from a hall by a free-standing closet and enclosure for the refrigerator and oven, **Figure 18–4.** This closet and enclosure cannot be recognized as free-standing (meaning they do not reach the ceiling) on the floor plan. However, this can be seen on Section 2/4. On the north side of the kitchen is a door to the deck outside. The east side of the hall has a

railing that continues up another set of stairs to the upper hall. Beyond this railing the floor is open to the playroom below.

Above the open area of the playroom is a loft that provides storage or extra sleeping space, **Figure 18–5.** Access to the loft is by a ladder in the upper hall. The loft is shown on a separate plan on Sheet 4. The loft is suspended on two beams built up of three 2 × 10s each. The north and south sides of the loft are enclosed by a wall 2′ – 6″ high. The east and west ends have a railing.

One of the doors in the upper hall opens into bedroom #1. On the east side of this bedroom is a hall that leads past a large (7' – 1" × 5' – 1") closet to toilet #1 in the northeast corner of the house. This toilet room has only a lavatory and water closet. A tub in the next room also serves toilet #2 to the east. Toilet #2 has a closet with shelves for linen storage. This toilet room can also be entered from the upper hall. The remaining room is bedroom #2.

The garage is attached to the house only by the wood deck and the roof. The garage is a rectangular building with 4-inch walls, an overhead door, and a walk-through door.

## Finding Dimensions

When you understand the relationships of the rooms to one another, you are ready to look for more detailed information. Frame walls are dimensioned in one of three ways, **Figure 18–6.** Exterior walls are usually dimensioned to the face of the studs or the face of the sheathing. Interior walls may be dimensioned either to the face of the studs or to their centerlines.

When walls are dimensioned to their centerlines, one-half of the wall thickness must be subtracted to find the face of the studs. For example, in **Figure 18–7,** the end walls are 12' – 4" on centers. However, the plates for the side walls are 12' – 0½" long (12' – 4" minus 3½", the width of the studs, equals 12' – 0½").

Dimensions are usually given in a continuous string where practical, **Figure 18–8.** Overall dimensions and major wall locations are given outside the view. Minor partitions and more detailed features are dimensioned either on or off the view, whichever is most practical.

The Lake House includes an angled wall. The length of such walls can be accurately found by trigonometry. However, the accuracy obtained by measuring with an architect's scale should be adequate for normal estimating.

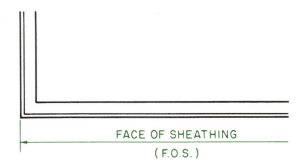

FACE OF SHEATHING
( F.O.S. )

FACE OF FRAMING
( F.O.F. )

ON CENTERS
( O.C. )

**Figure 18–6.** Three types of dimensioning.

12'-4" O.C.

2 X 4 FRAMING

12'-0½" F.O.F.

**Figure 18–7.** It is important to consider the true size of the framing material when working with O.C. dimensions.

**Figure 18–8.** Dimensions are placed in neat strings as much as possible. *Courtesy of Robert C. Kurzon.*

As each wall is laid out, the plates are cut to length, and the positions of all openings and intersecting walls are marked. The wall frame is usually assembled flat on the deck. Its position is marked on the deck with a chalkline; then the assembly is tipped up and slid into place.

Commercial buildings may require a thicker wall, where the electrical service panel is flush mounted and the panel box is deeper than the standard 3½-inch wall.

Many times this thicker wall requirement is not shown on the floor plan or drawing details.

The utilities that are to be stubbed up from a concrete floor slab into a wall are installed prior to floor slab-concrete placement and, therefore, prior to the wall locations being determined by the framing contractor. The utility installer must be able to accurately predetermine the specific wall locations.

## ✓ CHECK YOUR PROGRESS

Can you perform these tasks?

- List dimensions for locating all the walls and partitions in a house.

- Explain whether wall dimensions are to the face of sheathing, to the face of framing, or on centers.

- Identify plumbing walls shown on floor plans.

## ASSIGNMENTS

Refer to the Lake House drawings (in the packet) to complete the following assignment.

1. What are the inside dimensions of each room? Disregard slight irregularities in room shape, but remember to allow for wall thicknesses.

| Room | N-S dimension | E-W dimension |
|------|---------------|---------------|
| Utility | X | |
| Basement Bath | X | |
| Living Room | X | |
| Bedroom #1 | X | |
| Bedroom #2 | X | |
| Toilet #1 | X | |
| Tub | X | |
| Toilet #2 | X | |
| Loft | X | |

2. How many lineal feet of 2 × 4 frame wall are there in the basement?

3. How many lineal feet of 2 × 6 frame wall are there in the basement?

4. What is between the dining room and the living room?

5. What is directly below the loft?

6. What is directly below the tub room?

7. How many lineal feet of 2 × 6 frame wall are there on the Upper Level Floor Plan?

8. What utility must be roughed in prior to concrete-slab placement?

9. Where would you find the utility concrete-slab rough-in requirements?

# Framing Openings in Walls

## Objectives

After completing this unit, you will be able to perform the following tasks:

○ Describe typical rough openings.

○ Locate and interpret specific information for framing openings.

Two types of dimensions must be known before window and door openings can be framed: location and size. Opening locations are given by dimensioning to their centerlines in a string of dimensions outside the floor plan, **Figure 19–1.** Such dimensions are usually given from the face of the studs. One-half the rough opening is then allowed on each side of the centerline.

**Figure 19–1.** The locations of openings in framed walls are usually given to their centerlines. The ± on the sliding door dimensions allows the builder to place the doors next to the corner post. *Courtesy of Robert C. Kurzon.*

# Dimensions of Rough Openings

The size of the *rough opening* (R.O.), or opening in the framing, is listed on the door and window manufacturers' specifications. Door and window schedules are lists of all doors and windows in the house, usually with nominal dimensions, **Figure 19–2**. Doors and windows are identified on the floor plan by a *mark*—a letter or number. All doors or windows of a certain size and type have the same mark. Each mark is listed on the schedule with the information for the doors or windows of that type and size.

## SCHEDULE OF WINDOWS

| ⬡ | SIZE | TYPE | GLASS | COMMENTS |
|---|------|------|-------|----------|
| A | 3'-0" × 3'-0" | THERMAL BREAK ALUMINUM | 3/8" INSULATED | |
| B | 3'-0" × 5'-0" | D.O. | D.O. | |
| C | 4'-0" × 4'-0" | D.O. | D.O | |
| D | 4'-0" × 5'-0" | D.O. | D.O. | |
| E | 5'-0" × 5'-0" | D.O. | D.O. | |
| F | 6'-0" × 4'-0" | D.O | 1" INSULATED | |
| G | 6'-0" × 5'-0" | D.O. | 1" INSULATED | |
| H | 6'-0" × 6'-8" | D.O. | 1" TEMP. INSUL. | GLIDING GLASS DOORS |
| | | | | |

## SCHEDULE OF DOORS

| ◯ | SIZE | | THK | DOOR MAT'L | DOOR TYPE | FRAME MAT'L | FRAME TYPE | GLASS | LOCKSET | COMMENTS |
|---|------|---|-----|------------|-----------|-------------|------------|-------|---------|----------|
| 001 | 2'-8" × 6'-8" | | 1 3/8 | WD | C | WD. | A | | CLASSROOM | |
| 002 | 2'-0" × 6'-8" | | 1 3/8 | WD | C | WD. | A | | PRIVACY | |
| 003 | PR 2'-0" × 6'-8" | | 1 3/8 | WD | D | WD. | A | | PASSAGE R.H. DUMMY TRIM L.H. | |
| | | | | | | | | | | |
| 101 | 3'-0" × 6'-8" | | 1 3/4 | H.M. | A | WD. | A | 1" INSUL. | ENTRY | 1'-8" SIDELIGHT |
| 102 | 3'-0" × 6'-8" | | 1 3/4 | H.M. | B | WD. | A | | ENTRY | |
| 103 | PR 2'-4" × 6'-8" | | 1 3/8 | WD | D | WD. | B | | NONE | BI-PASSING HDWARE |
| 104 | 2'-6" × 6'-8" | | 1 3/8 | WD. | C | WD. | B | | PRIVACY | |
| 105 | PR 2'-6" × 6'-8" | | 1 3/8 | WD | D | WD. | B | | PASSAGE R.H. DUMMY TRIM L.H. | |
| 106 | 2'-0" × 6'-8" | | 1 3/8 | WD | C | WD. | B | | PRIVACY | |
| 107 | (4) 1'-0" × 6'-8" | | 1 3/8 | WD | E | WD. | B | | NONE | BI-FOLDING HDWARE |
| 108 | 2'-0" × 6'-8" | | 1 3/8 | WD | C | WD. | B | | PRIVACY | |
| 109 | 2'-0" × 6'-8" | | 1 3/8 | WD | C | WD. | B | | PRIVACY | |
| 110 | 2'-6" × 6'-8" | | 1 3/8 | WD | C | WD. | B | | PRIVACY | |
| 111 | 2'-0" × 6'-8" | | 1 3/8 | WD | C | WD. | B | | PASSAGE | |
| 112 | 2'-6" × 6'-8" | | 1 3/8 | WD | C | WD. | B | | PRIVACY | |
| | | | | | | | | | | |
| | | | | | | | | | | |

**Figure 19–2.** Window and door schedules list all the windows and doors and their sizes. *Courtesy of Robert C. Kurzon.*

**Andersen®** WINDOWS·DOORS

## Table of Basic Casement Unit Sizes  Scale 1/8" = 1'-0" (1:96)

| Unit Dimension | 1'-5" (432) | 1'-8 1/2" (521) | 2'-0 1/8" (613) | 2'-4 3/8" (721) | 2'-7 1/2" (800) | 2'-11 15/16" (913) | 2'-9 3/4" (857) | 3'-4 3/4" (1035) | 4'-0" (1219) | 4'-8 1/2" (1435) |
|---|---|---|---|---|---|---|---|---|---|---|
| **Minimum Rough Opening** | 1'-5 1/2" (445) | 1'-9" (533) | 2'-0 5/8" (625) | 2'-4 7/8" (733) | 2'-8" (813) | 3'-0 1/2" (927) | 2'-10 1/4" (870) | 3'-5 1/4" (1048) | 4'-0 1/2" (1232) | 4'-9" (1448) |
| Unobstructed Glass* | 12 5/8" (321) | 16 1/8" (410) | 19 3/4" (502) | 24" (610) | 27 1/8" (689) | 31 9/16" (802) | 12 5/8" (321) | 16 1/8" (410) | 19 3/4" (502) | 24" (610) |
| Unobstructed Glass Transom Units Only | 12 3/16" (310) | 15 11/16" (398) | 19 5/16" (491) | 23 9/16" (599) | 26 11/16" (678) | 31 1/8" (791) | 28 15/16" (735) | 35 15/16" (913) | 43 3/16" (1097) | 51 11/16" (1313) |

| 1'-0" (305) / 1'-0 1/2" (318) / 7 3/16" (183) | CTR1510 ‡ | CTR1810 ‡ | CTR2010 ‡ | CTR2410 ‡ | CTR2810 ‡ | CTR3010 ‡ | CTR2910 ‡ | CTR3410 ‡ | CTR4010 ‡ | CTR4810 ‡ |

CTR units are non-venting.

| 2'-0 1/8" (613) / 2'-0 5/8" (625) / 19 5/16" (491) | CR12 | CN12 | C12 | CW12 | | | | | | |
| 2'-4 3/8" (721) / 2'-4 7/8" (733) / 23 9/16" (598) | CR125 | CN125 | C125 | CW125 | CX125 | | | | | |
| 2'-11 15/16" (913) / 3'-0 1/2" (927) / 31 1/8" (791) | CR13 | CN13 | C13 | CW13 | CX13 | CXW13 | CR23 | CN23 | C23 | CW23 |
| 3'-4 13/16" (1037) / 3'-5 3/8" (1051) / 36" (914) | CR135 | CN135 | C135 | CW135†♦ | CX135***♦ | CXW135♦ | CR235 | CN235 | C235 | CW235†♦ |
| 4'-0" (1219) / 4'-0 1/2" (1232) / 43 3/16" (1097) | CR14 | CN14 | C14 | CW14†♦ | CX14♦ | CXW14♦ | CR24 | CN24 | C24 | CW24†♦ |
| 4'-4 13/16" (1341) / 4'-5 3/8" (1356) / 48" (1219) | CR145 | CN145 | C145 | CW145†♦ | CX145♦ | CXW145♦ | CR245 | CN245 | C245 | CW245†♦ |
| 4'-11 7/8" (1521) / 5'-0 3/8" (1534) / 55 1/16" (1399) | CR15 | CN15 | C15 | CW15†♦ | CX15♦ | CXW15***♦ | CR25 | CN25 | C25 | CW25†♦ |
| 5'-4 13/16" (1646) / 5'-5 3/8" (1660) / 60" (1524) | CR155 | CN155 | C155 | CW155†♦ | CX155♦ | CXW155***♦ | CR255 | CN255 | C255 | CW255†♦ |
| 5'-11 7/8" (1826) / 6'-0 3/8" (1838) / 67 1/16" (1703) | CR16 | CN16 | C16 | CW16†♦ | CX16♦ | CXW16**♦ | CR26 | CN26 | C26 | CW26†♦ |

* "Unobstructed Glass" measurement is for single sash only.
** These units have straight arm operators, see opening specifications.
† CW series units (except CW2, CW25 and CW3 height) open to 20" clear opening width using sill hinge control bracket. Bracket can be pivoted allowing for cleaning position. CW series units are also available with a 22" clear opening width.
‡ Andersen® art glass panels are available for these units by special order only. Contact your Andersen® supplier. Panels are available for all other units on this page through normal ordering process.
♦ These units meet or exceed the following dimensions: Clear Openable Area of 5.7 sq. ft., Clear Openable Width of 20" and Clear Openable Height of 24", when appropriate hardware (straight arm or split arm) is specified.
• Casement transom units (CTR) may be rotated to be used as a casement or awning sidelight.
• Rough opening dimensions may need to be increased to allow for use of building wraps, flashings, sill panning, brackets, fasteners or other items. See page 10 for more details.
• "Unit Dimension" always refers to outside frame to frame dimension.
• Dimensions in parentheses are in millimeters.
• When ordering, be sure to specify color desired: White, Sandtone, Terratone® or Forest Green.

Left   Right   Stationary

**Venting Configuration**

Hinging shown on size table is standard. Specify left, right or stationary, as viewed from the outside. For other hinging of multiple units, contact your local supplier.

**Figure 19–3.** Typical page from a window catalog. *Courtesy of Iroquois Millwork Corporation.*

The rough opening dimensions are not usually given on the drawings. They should be obtained from the manufacturer. Window manufacturers do not list the same size information in their catalogs, but they do list the rough opening size. Other sizes they sometimes list are nominal size or unit dimensions (approximate overall size of the window) and glass size (size of the actual glass in the window), **Figure 19–3**. If the finished doorway, the **jamb,** is to be built on the site by a carpenter, the rough opening size will not be available. In this case the rough opening for swing doors can be built 2 inches wider and 2 inches higher than the door. If the door is another type, you must first determine what the finished opening is to be. Manufacturers of hardware for sliding and bifold doors specify rough opening sizes for doors installed with their hardware. These manufacturer's specifications should be consulted as the opening is being framed.

Sizes of doors and windows are given with the width first and height second. To further simplify dimensioning, they are often listed as feet/inches. For example, a $2^6 \times 6^8$ door is 2 feet 6 inches wide by 6 feet 8 inches high.

The information in this unit is intended only to help you determine rough opening sizes. Electrical utility installers should familiarize themselves with the locations and sizes of these rough openings prior to installing the electrical wiring. More information about windows and doors can be found in Unit 25.

## Framing Openings

In stud-wall construction, it is usually necessary to cut off or completely eliminate one or more studs where windows and doors are installed. The load normally carried by these studs must be transferred to the sides of the opening. The construction over an opening that transfers this load to the sides of the opening is called the *header.* Additional studs are installed at the sides to carry this load, **Figure 19–4.**

A set of construction drawings includes details showing the type of rough opening construction intended. The simplest type of wood construction uses two 2 × 10s or 2 × 12s with plywood spacers between to form the header, **Figure 19–5.** A flat 2 × 4 may be nailed to the bottom of the header to reduce the height of the top of the opening, but this flat 2 × 4 is not considered as part of the size of the header.

**Figure 19–4.** The jack studs support the header.

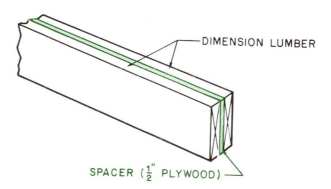

**Figure 19–5.** Solid wood header.

To conserve lumber in nonbearing walls, the header may be two 2 × 4s. The area over the header is framed with cripple studs. The area below a window also is framed with *cripple studs,* **Figure 19–6.** These cripple studs are installed on the normal spacing for wall studs. This may require a cripple within a few inches of a side jack stud, but is necessary to provide a nailing surface for the sheathing.

Pocket doors are sliding doors that slide into an opening in the wall, thereby allowing the largest possible opening without swinging into the room. Most builders install pocket doors as a preassembled frame, **Figure 19–7.** The wall framing to provide the opening into which a pocket door assembly can be installed is

Figure 19–6. The spaces above and below the window opening are framed with cripple studs.

generally the same as the framing for any other type of door. Pocket door manufacturers provide rough opening dimensions for their products.

Several systems of metal framing are available for light construction. The basic elements of light-gauge metal framing are the same as for wood framing, **Figure 19–8.** These systems use top and bottom wall plates, studs, and floor joists. To make light-gauge framing compatible with wood, the metal members are made in common sizes for wood framing. The greatest difference is that metal framing is joined with screws instead of nails.

Headers over openings in light-gauge metal framing are usually very similar to those in wood framing. However, the system designed by the metal framing manufacturer should always be followed.

Figure 19–7. Pocket door frame. *Courtesy L. E. Johnson Products, Inc.*

**Figure 19–8.** The parts of this steel frame are the same as those of a wood frame. Notice the window header and cripple studs. *Courtesy of Zinc Institute, Inc.*

## ✓ CHECK YOUR PROGRESS

- Identify each door or window listed on a schedule of doors and windows according to its location on the floor plan.

- Determine the rough opening for a window from the information in a manufacturer's catalog.

- Describe the framing for a window opening.

- Describe the framing for a door opening.

- Determine the rough opening for a door.

- Explain the difference between a solid wood header and a trussed header.

## ASSIGNMENT

Refer to the Lake House drawings (in the packet) to complete this assignment.

1. What type of header should be used over the door from the hall to bedroom #1?

2. What type of header should be used over the door from the deck to the kitchen?

3. Why should these two headers be made differently?

4. What is the length of the header over the garage overhead door? Allow for two trimmers at each side and 1¼" for jambs at each side.

5. What are the R.O. dimensions for the door from the deck into the garage?

6. Name the location and give the R.O. dimensions for each interior door on the Upper Level Floor Plan.

7. According to **Figure 19–3,** what are the R.O. dimensions for the windows in bedroom #2?

8. How long is the header over the window in bedroom #2?

9. How many cripple studs are needed beneath the windows in the south wall of the living room?

10. What is the length of the cripple studs beneath the window in bedroom #2? (Assume the bottom of the header is 6'-8½" from the top of the subfloor.)

# UNIT 20

# Roof Construction Terms

## Objectives

After completing this unit, you will be able to perform the following tasks:

O Identify common roof types.

O Define the terms used in laying out and constructing a roof.

O State roof pitches as a ratio of rise to run or as a fraction.

## Types of Roofs

Several types of roofs are commonly used in residential construction, **Figure 20–1.** Variations of these roof types may be used to create certain architectural styles.

The *gable roof* is one of the most common types used on houses. The gable roof consists of two sloping sides that meet at the ridge. The triangle formed at the ends of the house between the top plates of the wall and roof is called the **gable.**

The *gambrel roof* is similar to the gable roof. On this roof, the sides slope very steeply from the walls to a point about halfway up the roof. Above this point, they have a more gradual slope.

The *hip roof* slopes on all four sides. The hip roof has no exposed wall above the top plates. This means that all four sides of the house are equally protected from the weather.

The *mansard roof* is similar to the hip roof, except the lower half of the roof has a very steep slope and the top half is more gradual. This roof style is used extensively in commercial construction—on stores, for example.

The *shed roof* is a simple sloped roof with no ridge. A shed roof is much like one side of a gable roof. This type of roof is used in modern architecture and for additions to existing buildings.

## Roof Construction Terms

The roof construction terms used with trussed roofs, the most common type of roof framing for light construction, are defined below and shown in **Figure 20–2.**

O **Top chord** is the top member in the truss.

O **Bottom chord** is the bottom member of the truss and acts as a ceiling joist for the space below.

O A **web** member is any of the interior bracing between the top chord and the bottom chord.

O A **gusset** is a reinforcing piece of metal or plywood fastened to the truss, where members are joined.

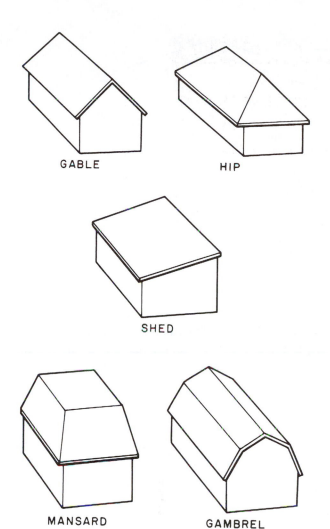

Figure 20–1. Common roof types.

GABLE

HIP

SHED

MANSARD

GAMBREL

○ **Span** is the distance between the outsides of the walls covered by a roof.

○ **Rise** is the vertical distance or height of the roof.

○ The **overhang** is the horizontal distance covered by the roof outside the walls.

○ The **tail** is the portion of the top chord that is outside the walls.

○ The **pitch** of a roof is a way of indicating how steep the roof is. The pitch of a roof is usually given as the number of inches of rise per foot of run. This is called *slope* or *unit rise*.

In writing unit rise, the slope is given as the number of inches of rise for every 12 inches of run, **Figure 20–3.** The rise per foot of run is given on drawings with the symbol ◪. The horizontal leg of this triangle represents the run. The vertical leg represents the rise, **Figure 20–4.** Notice that **Figure 20–4** shows two roof pitches. These pitches are written as 10 in 12 and 5 in 12.

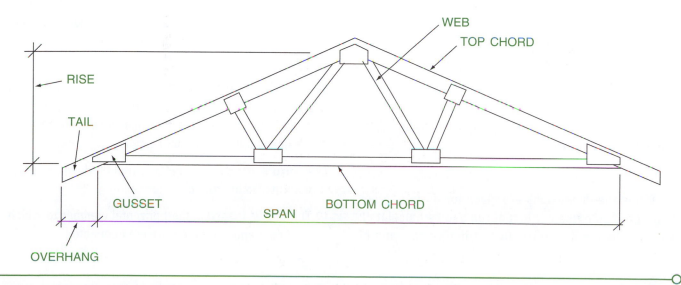

RISE

TAIL

GUSSET

OVERHANG

WEB

TOP CHORD

SPAN

BOTTOM CHORD

Figure 20–2. Parts of a roof truss.

**Figure 20–3.** Rise per foot of run. This roof has a 4 in 12 slope.

**Figure 20–4.** Roof slope is usually shown on building elevations. *Courtesy of Home Planners, Inc.*

When the dimensions are given for a run of other than 12 feet, the rise per foot of run can be calculated: Divide 12 by the actual run to find the proper ratio of

rise. Multiply this result by the actual rise to find the rise per foot of run. For example, if the run is 14′ – 6″ and the rise is 5′ – 0″; what is the rise per foot of run? 12 = 14′ – 6″ (14.5′) = 0.83 ÷ 5 = 4.15. The result (4.15) is close enough to 4 that rafter calculations can be based on 4 in 12.

Although most modern roof frames are built with **trusses,** it is valuable to know something about rafter-framed roofs, because some special situations call for **rafters** to be cut and installed on the site. The terms defined in the following list are illustrated in **Figure 20–5.**

O **Run** is the vertical distance covered by one rafter. Run does not include any part of the rafter that extends beyond the wall. On a common two-sided roof with both sides having the same pitch, the run is one-half the span.

O The **measuring line** is an imaginary line along which all roof dimensions are taken. The measuring line of a rafter is a line parallel to its edges and passing through the deepest part of the bird's mouth.

O A **bird's mouth** is a notch cut in the lower edge of the rafter to fit around the top of the wall.

O The **ridge board** is the horizontal member to which the upper ends of the rafters are connected.

**Figure 20–5.** Common roof terms.

✓ CHECK YOUR PROGRESS

Can you perform these tasks?

● Identify the following roof types from information given on construction drawings: gable, gambrel, hip, mansard, and shed.

● List each of the following from information given on construction drawings: span, run, rise, overhang, rafter tails, and slope.

ASSIGNMENT

1. Give the following information for the roof shown in **Figure 20–6:**
   a. Span
   b. Run
   c. Rise
   d. Overhang
   e. Length of the rafter tails
   f. Unit rise

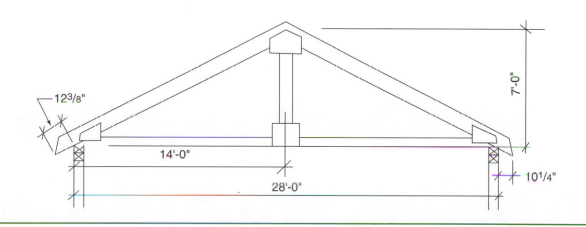

**Figure 20–6.**

Refer to the Lake House drawings (in the packet) to complete the rest of the assignment.

2. What style roof is used for most of the Lake House?

3. What is the span of the rafters over the garage?

4. What is the rise per foot of run of the rafters over the garage?

5. What is the run of the rafters over bedroom #1?

6. What is the rise per foot of run of the rafters over bedroom #1?

7. What is the overhang of the rafters over bedroom #1?

# Roof Trusses

## Truss Information on Drawings

The roof is designed to support weight, withstand weather, and give protection to the spaces below. In most roofs the frame is made up primarily of trusses; however, some special framing situations still call for rafters. Designing trusses to withstand all the stresses that can be placed on them by wind or snow, or both, requires the skills of an engineer. Generally, the architect or designer who designs the house has some idea of what size and type of trusses will be required. The plans and elevations for the building show a lot of information about the trusses. The pitch of the roof can be seen in the building elevations, **Figure 21–1.** Section views through the building show trusses, but the truss shown in a section might not be the truss the engineer designs. Notice that a note near the truss in **Figure 21–2** says "NYS P.E. Approved Roof Trusses @24" OC." This note indicates that the trusses will be designed by a professional engineer (P.E.) who is licensed in New York State (NYS) and that they will be spaced 24 inches from the centerline of one truss to the centerline of the next truss. You will learn in a later discussion that the actual truss specified by the engineer is not quite like the one on the building section.

## Truss Engineering Drawings

To see what the engineered trusses are, refer to the **truss drawings** from the engineer. These usually include a **delivery sheet** showing all the types of trusses used in the building, **Figure 21–3.** This sheet usually shows the following information:

*Profile:* A small drawing showing the basic shape of the truss.

*Quantity:* Number of each type of truss needed for the building.

*Truss ID:* The engineer's letter/number designation for that truss type.

*Span:* The distance between the outside walls the truss will cover.

*Truss type:* Some trusses are referred to as *piggyback trusses.* Large trusses can be too large to transport to the construction site, so they are delivered in two parts. The top of the main truss is flat, so the piggyback truss can be applied to it after it is in place.

*Slope or pitch:* The example here shows slope or the unit rise of the roof.

## Objectives

After completing this unit, you will be able to perform the following tasks:

O Find information about trusses on plans, elevations, and sections.

O Identify the appropriate truss engineering drawing for each truss in a roof frame.

O Find key information on a truss engineering drawing.

**Figure 21–1.** Roof pitch can be seen in building elevations. *Courtesy of Robert Marini Builders, Inc.*

---

*LOH:* Left overhang—the amount of the top chord that must extend beyond the building wall.

*ROH:* Right overhang—the amount of the top chord that must extend beyond the building wall.

The packet of engineering drawings usually includes a **truss layout plan, Figure 21–4.** This plan shows each truss in place on the building, with a label to indicate the truss ID. If you know where in the building the section view in **Figure 21–2** is taken from, you will be able to find that truss on the truss layout. The section view is labeled as A/6A, so look at the Second Floor Plan, **Figure 21–5,** to see where the cut was made to take Section A. The section was cut through the foyer, which is on the left

side of the house. Looking again at the section, you can see the rafters for the **dormer,** which is above the main entrance to the house. In the truss layout, **Figure 21–4,** a single-line triangle indicates where the dormer will be. A notation above that triangle indicates that the eight trusses in the dormer area are T02A. This is the fourth truss down from the top on the delivery sheet shown in **Figure 21–3.**

The truss engineering drawings include a detail sheet for each truss type. Truss T02A is shown in **Figure 21–6.** The **truss detail** includes a drawing of the truss and all the engineering data required to build the truss. Carpenters do not usually build trusses, but the same packet of engineering drawings that go to the site with

**Figure 21–2.** Building section. *Courtesy of Robert Marini Builders, Inc.*

the trusses is also used in the truss shop to build the trusses. It is very important for the carpenter to read all the information on the detail even though it might not all be needed to set the trusses on the building. For example, in some applications, the engineer will call for the trusses to be applied in 2-ply or 3-ply. If the trusses are plied, they are fastened together, face to face, so that all the truss members are joined in two or three layers. In this case the ply is shown as "1" in the box at the top of the detail sheet.

## Gable Ends

A roof gable is the triangle formed at the end of the roof by the two top chords of the truss and its bottom chord. The roof framing at a gable end must include studs, so there will be a place to attach whatever sheathing and siding are to be applied. A gable-end truss is used for this. When an overhang is needed on the gable end, it can be built as part of the gable-end truss, **Figure 21–7.**

**Bellevue Builders Supply**
500 Duanesburg Rd
Schenectady, NY 12306
Phone: (518)355-7190  Fax: (518)355-1371

| Project: | Block No: |
| Model: | Lot No: |

| Contact: | Site: | Office: |
| Name: | | |
| Phone: | | |
| Fax: | | |

Tentative Delivery Date:

To:
MARINI BLDRS.

Deliver To:
BELMONT MODEL A3635

# Delivery List

| Job Number: | |
| Page: | 1 |
| Date: | 01-29-2004 - 3:36:11 PM |
| Project ID: | A3635 |

| Account No: | |
| Designer: | WTC |
| Salesperson: | |
| Quote Number: | |

| Profile: | Qty: | Truss Id: | Span: | Truss Type: | Slope: | LOH | ROH | Load By: |
|---|---|---|---|---|---|---|---|---|
| | 11 | T01<br>136 lbs. each | 31 - 9 - 0 | ROOF TRUSS | 8.00<br>0.00 | 0 - 0 - 0 | 0 - 10 - 8 | |
| | 1 | T01GE<br>173 lbs. each | 31 - 9 - 0 | ROOF TRUSS<br>3 Rows Lat Brace | 8.00<br>0.00 | 0 - 0 - 0 | 0 - 10 - 8 | |
| | 13 | T02<br>176 lbs. each | 37 - 0 - 0 | PIGGYBACK TRU<br>1 Row Lat Brace | 8.00<br>0.00 | 0 - 10 - 8 | 1 - 2 - 8 | |
| | 8 | T02A<br>174 lbs. each | 37 - 0 - 0 | PIGGYBACK TRU<br>1 Row Lat Brace | 8.00<br>0.00 | 0 - 10 - 8 | 0 - 0 - 0 | |
| | 1 | T02GE<br>209 lbs. each | 37 - 0 - 0 | PIGGYBACK TRU<br>7 Rows Lat Brace | 8.00<br>0.00 | 0 - 10 - 8 | 1 - 2 - 8 | |
| | 1 | T02GES<br>210 lbs. each | 37 - 0 - 0 | PIGGYBACK TRU<br>1 Row Lat Brace | 8.00<br>0.00 | 0 - 10 - 8 | 1 - 2 - 8 | |
| | 3 | T03<br>93 lbs. each | 22 - 0 - 0 | ROOF TRUSS | 10.00<br>0.00 | 0 - 10 - 8 | 1 - 2 - 8 | |
| | 5 | T03A<br>90 lbs. each | 22 - 0 - 0 | ROOF TRUSS | 10.00<br>0.00 | 0 - 0 - 0 | 0 - 0 - 0 | |
| | 1 | T03B<br>89 lbs. each | 21 - 8 - 8 | ROOF TRUSS | 10.00<br>0.00 | 0 - 0 - 0 | 0 - 0 - 0 | |
| | 1 | T03GE<br>114 lbs. each | 22 - 0 - 0 | ROOF TRUSS | 10.00<br>0.00 | 0 - 10 - 8 | 1 - 2 - 8 | |
| | 21 | T2PB<br>30 lbs. each | 11 - 5 - 8 | PIGGYBACK | 8.00<br>0.00 | 0 - 0 - 0 | 0 - 0 - 0 | |
| | 2 | T2PBGE<br>34 lbs. each | 11 - 5 - 8 | PIGGYBACK | 8.00<br>0.00 | 0 - 0 - 0 | 0 - 0 - 0 | |
| | 1 | VM01<br>63 lbs. each | 15 - 5 - 10 | ROOF TRUSS | 10.00<br>0.00 | 0 - 0 - 0 | 0 - 0 - 0 | |

**Figure 21–3.** Truss delivery list. *Courtesy of Robert Marini Builders, Inc.*

**Figure 21–4.** Truss plan. *Courtesy of Robert Marini Builders, Inc.*

**Figure 21–5.** Second Floor Plan. *Courtesy of Robert Marini Builders, Inc.*

| Job | Truss | Truss Type | Qty | Ply | BELLEVUE BUILDERS SUPPLY |
|---|---|---|---|---|---|
| A3635 | T02A | PIGGYBACK TRUSS | 8 | 1 | MARINI BLDRS. BELMONT MODEL |
| | | | | | Job Reference (optional) |

Bellevue Builders Supply, Schenectady, NY 12306      5.200 s Aug 19 2003 MiTek Industries, Inc.  Thu Jan 29 15:29:03 2004  Page 1

Scale = 1:83.7

Plate Offsets (X,Y): [2:1-8-2,0-2-0], [2:0-0-12,Edge], [7:0-3-4,0-2-12], [9:0-4-4,0-3-0], [12:0-0-7,Edge]

| LOADING (psf) | | SPACING | 2-0-0 | CSI | | DEFL | in | (loc) | l/defl | L/d | PLATES | GRIP |
|---|---|---|---|---|---|---|---|---|---|---|---|---|
| TCLL | 50.0 | Plates Increase | 1.15 | TC | 0.88 | Vert(LL) | -0.47 | 15-17 | >942 | 360 | MII20 | 169/123 |
| TCDL | 10.0 | Lumber Increase | 1.15 | BC | 1.00 | Vert(TL) | -0.61 | 15-17 | >717 | 180 | MII20H | 127/93 |
| BCLL | 0.0 * | Rep Stress Incr | YES | WB | 0.58 | Horz(TL) | 0.17 | 12 | n/a | n/a | | |
| BCDL | 10.0 | Code | IBC2000/ANSI95 | (Matrix) | | | | | | | Weight: 175 lb | |

**LUMBER**
TOP CHORD   2 X 4 SPF-S 1650F 1.5E *Except*
     T3 2 X 4 SPF-S 2100F 1.8E, T4 2 X 4 SPF-S 2100F 1.8E
BOT CHORD   2 X 4 SPF-S 1650F 1.5E *Except*
     B2 2 X 4 SPF-S 2100F 1.8E
WEBS      2 X 4 SPF No.2 *Except*
     W7 2 X 4 SPF Stud, W6 2 X 4 SPF Stud, W3 2 X 4 SPF Stud
SLIDER      Left 2 X 6 SPF-S 1650F 1.5E 2-8-0, Right 2 X 6 SPF-S 1650F 1.5E 3-8-6

**BRACING**
TOP CHORD   Sheathed or 2-4-6 oc purlins, except
     2-0-0 oc purlins (3-7-2 max.): 7-9.
BOT CHORD   Rigid ceiling directly applied or 10-0-0 oc bracing.
WEBS      1 Row at midpt      8-15

**REACTIONS** (lb/size)   12=2797/0-5-8, 2=2892/0-5-8
     Max Horz 2=295(load case 7)
     Max Uplift12=-256(load case 9), 2=-318(load case 8)
     Max Grav 12=3381(load case 3), 2=3548(load case 2)

**FORCES** (lb) - Maximum Compression/Maximum Tension
TOP CHORD   1-2=0/81, 2-3=-5217/632, 3-4=-4856/656, 4-5=-4601/599, 5-6=-4381/616, 6-7=-4534/745, 7-8=-3388/611, 8-9=-3388/611,
     9-10=-4471/665, 10-11=-4769/629, 11-12=-5007/605
BOT CHORD   2-17=-491/4119, 17-18=-356/3140, 16-18=-356/3140, 15-16=-356/3140, 14-15=-228/3136, 14-19=-228/3136,
     19-20=-228/3136, 13-20=-228/3136, 13-21=-402/3864, 21-22=-402/3864, 12-22=-402/3864
WEBS      4-17=-577/204, 6-17=-675/223, 7-17=-265/1385, 7-15=-223/1212, 8-15=-1031/267, 9-15=-235/1256, 9-13=-155/957,
     10-13=-687/278

**NOTES**
1) Wind: ASCE 7-98; 90mph; h=30ft; TCDL=5.0psf; BCDL=5.0psf; Category II; Exp C; enclosed; MWFRS gable end zone and C-C
     Exterior(2) zone; cantilever left and right exposed ; end vertical left and right exposed; Lumber DOL=1.33 plate grip DOL=1.33. This truss
     is designed for C-C for members and forces, and for MWFRS for reactions specified.
2) TCLL: ASCE 7-98; Pf=50.0 psf (roof snow); Exp C; Partially Exp.
3) Unbalanced snow loads have been considered for this design.
4) This truss has been designed for 1.00 times flat roof load of 50.0 psf on overhangs non-concurrent with other live loads.
5) Provide adequate drainage to prevent water ponding.
6) This truss has been designed for a 10.0 psf bottom chord live load nonconcurrent with any other live loads per Table 1607.1 of IBC-00.
7) All plates are MII20 plates unless otherwise indicated.
8) * This truss has been designed for a live load of 20.0psf on the bottom chord in all areas where a rectangle 3-6-0 tall by 1-0-0 wide will fit
     between the bottom chord and any other members.
9) Provide mechanical connection (by others) of truss to bearing plate capable of withstanding 256 lb uplift at joint 12 and 318 lb uplift at
     joint 2.
10) Design assumes 4x2 (flat orientation) purlins at oc spacing indicated, fastened to truss TC w/ 2-10d nails.

**LOAD CASE(S)** Standard

**Figure 21–6.** T02A engineering detail. *Courtesy of Robert Marini Builders, Inc.*

**Figure 21–7.** Gable-end truss with overhang. *Reprinted with permission from the Wood Truss Council of America (WTCA). For more information visit* www.woodtruss.com.

## ✓ CHECK YOUR PROGRESS

Can you perform these tasks?

● Find information about which way the trusses run by looking at a floor plan.

● Find information about the pitch or slope of a roof by looking at the building elevations.

● Find the number of trusses of any given type required for a roof.

● Identify the truss engineering detail for any given location in a roof plan.

## ASSIGNMENT

Refer to the figures in this unit and **Figures 21–8** and **21–9** to answer the following questions.

1. What is the slope of the garage roof?

2. What is the span of the trusses in the garage roof?

3. What are the truss IDs for the two types of trusses in the garage roof?

4. What is the difference between the two types of trusses in the garage roof?

5. Where does the T02GE truss go?

6. Where does the T2PBGE truss go?

7. What is the on-center spacing from T02GE to the next closest truss?

8. What is the slope of the roof over bedroom #2?

| Job | Truss | Truss Type | Qty | Ply | BELLEVUE BUILDERS SUPPLY |
|---|---|---|---|---|---|
| A3635 | T02GE | PIGGYBACK TRUSS | 1 | 1 | MARINI BLDRS. BELMONT MODEL |
| | | | | | Job Reference (optional) |

Bellevue Builders Supply, Schenectady, NY 12306      5.200 s Aug 19 2003 MiTek Industries, Inc.  Thu Jan 29 15:29:39 2004  Page 1

Scale = 1:71.9

Plate Offsets (X,Y): [2:0-2-10,0-1-8], [10:0-3-0,0-0-2], [16:0-4-4,0-2-4]

| LOADING (psf) | | SPACING | 2-0-0 | CSI | | DEFL | in | (loc) | l/defl | L/d | PLATES | GRIP |
|---|---|---|---|---|---|---|---|---|---|---|---|---|
| TCLL | 50.0 | Plates Increase | 1.15 | TC | 0.24 | Vert(LL) | n/a | - | n/a | 999 | MII20 | 169/123 |
| TCDL | 10.0 | Lumber Increase | 1.15 | BC | 0.09 | Vert(TL) | 0.01 | 24 | >999 | 180 | | |
| BCLL | 0.0 * | Rep Stress Incr | YES | WB | 0.39 | Horz(TL) | 0.00· | 23 | n/a | n/a | | |
| BCDL | 10.0 | Code IBC2000/ANSI95 | | (Matrix) | | | | | | | Weight: 209 lb | |

**LUMBER**
TOP CHORD  2 X 4 SPF-S 1650F 1.5E
BOT CHORD  2 X 4 SPF-S 1650F 1.5E
OTHERS  2 X 4 SPF Stud *Except*
   ST7 2 X 4 SPF No.2, ST7 2 X 4 SPF No.2, ST7 2 X 4 SPF No.2
   ST6 2 X 4 SPF No.2, ST7 2 X 4 SPF No.2, ST7 2 X 4 SPF No.2
   ST8 2 X 4 SPF No.2
WEDGE
Right: 2 X 4 SPF Stud

**BRACING**
TOP CHORD  Sheathed or 6-0-0 oc purlins.
BOT CHORD  Rigid ceiling directly applied or 10-0-0 oc bracing,  Except:
   6-0-0 oc bracing: 29-31.
WEBS  1 Row at midpt     13-34, 12-35, 11-36, 9-37, 14-33, 15-32,
                            16-31

**REACTIONS** (lb/size)  2=299/37-0-0, 23=363/37-0-0, 34=280/37-0-0, 35=281/37-0-0, 36=273/37-0-0, 37=248/37-0-0, 39=279/37-0-0,
   40=280/37-0-0, 41=284/37-0-0, 42=262/37-0-0, 43=338/37-0-0, 33=282/37-0-0, 32=269/37-0-0, 31=249/37-0-0,
   29=284/37-0-0, 28=279/37-0-0, 27=282/37-0-0, 26=272/37-0-0, 25=305/37-0-0
Max Horz 2=279(load case 8)
Max Uplift2=-62(load case 6), 23=-35(load case 7), 34=-61(load case 7), 35=-66(load case 6), 36=-52(load case 7),
   37=-18(load case 7), 39=-95(load case 8), 40=-84(load case 8), 41=-86(load case 8), 42=-83(load case 8),
   43=-99(load case 8), 33=-64(load case 6), 32=-63(load case 7), 31=-7(load case 6), 29=-88(load case 9),
   28=-86(load case 9), 27=-85(load case 9), 26=-85(load case 9), 25=-107(load case 9)
Max Grav 2=406(load case 2), 23=500(load case 3), 34=379(load case 2), 35=383(load case 3), 36=376(load case 3),
   37=346(load case 2), 39=379(load case 2), 40=380(load case 2), 41=385(load case 2), 42=357(load case 2),
   43=457(load case 2), 33=383(load case 3), 32=384(load case 2), 31=344(load case 3), 29=387(load case 3),
   28=379(load case 3), 27=382(load case 3), 26=370(load case 3), 25=411(load case 3)

**FORCES** (lb) - Maximum Compression/Maximum Tension
TOP CHORD  1-2=0/77, 2-3=-259/145, 3-4=-176/132, 4-5=-147/127, 5-6=-149/111, 6-7=-76/121, 7-8=-149/125, 8-9=-150/171,
   9-10=-141/170, 10-11=-52/176, 11-12=-52/176, 12-13=-52/176, 13-14=-52/176, 14-15=-52/176, 15-16=-54/175,
   16-17=-158/180, 17-18=-152/115, 18-19=-58/58, 19-20=-153/48, 20-21=-152/64, 21-22=-157/68, 22-23=-228/85,
   23-24=0/82
BOT CHORD  2-43=0/0, 42-43=0/0, 41-42=0/0, 40-41=0/0, 39-40=0/0, 38-39=0/0, 37-38=-0/0, 36-37=-0/0, 35-36=-0/0, 34-35=-0/0,
   33-34=-0/0, 32-33=-0/0, 31-32=-0/0, 30-31=-4/0, 29-30=-54/252, 28-29=-54/252, 27-28=-54/252, 26-27=-54/252,
   25-26=-54/252, 23-25=-54/252
WEBS  13-34=-339/81, 12-35=-344/86, 11-36=-335/73, 9-37=-315/33, 8-39=-342/113, 7-40=-340/104, 5-41=-343/106,
   4-42=-324/102, 3-43=-394/122, 14-33=-343/84, 15-32=-343/84, 16-31=-312/23, 17-29=-350/106, 18-28=-338/106,
   20-27=-342/105, 21-26=-330/102, 22-25=-372/135

**NOTES**
1) Wind: ASCE 7-98; 90mph; h=30ft; TCDL=5.0psf; BCDL=5.0psf; Category II; Exp C; enclosed; MWFRS gable end zone and C-C
   Exterior(2) zone; cantilever left and right exposed ; end vertical left and right exposed; Lumber DOL=1.33 plate grip DOL=1.33. This truss
   is designed for C-C for members and forces, and for MWFRS for reactions specified.
2) Truss designed for wind loads in the plane of the truss only.  For studs exposed to wind (normal to the face), see MiTek "Standard Gable
   End Detail"
3) TCLL: ASCE 7-98; Pf=50.0 psf (roof snow); Exp C; Partially Exp.
4) Unbalanced snow loads have been considered for this design.
5) This truss has been designed for 1.00 times flat roof load of 50.0 psf on overhangs non-concurrent with other live loads.
6) Provide adequate drainage to prevent water ponding.

Continued on page 2

**Figure 21–8.** T02GE engineering detail. *Courtesy of Robert Marini Builders, Inc.*

| Job | Truss | Truss Type | Qty | Ply | BELLEVUE BUILDERS SUPPLY |
|-----|-------|-----------|-----|-----|--------------------------|
| A3635 | T2PBGE | PIGGYBACK | 2 | 1 | MARINI BLDRS. BELMONT MODEL |
| | | | | | Job Reference (optional) |

Bellevue Builders Supply, Schenectady, NY 12306          5.200 s Aug 19 2003 MiTek Industries, Inc.  Thu Jan 29 15:31:41 2004  Page 1

Plate Offsets (X,Y): [5:0-3-3,0-0-12], [6:0-4-7,0-0-10], [8:0-2-0,0-0-12]

| LOADING (psf) | SPACING | 2-0-0 | CSI | | DEFL | in | (loc) | l/defl | L/d | PLATES | GRIP |
|---|---|---|---|---|---|---|---|---|---|---|---|
| TCLL 45.0 | Plates Increase | 1.15 | TC | 0.21 | Vert(LL) | n/a | - | n/a | 999 | MII20 | 169/123 |
| TCDL 10.0 | Lumber Increase | 1.15 | BC | 0.04 | Vert(TL) | n/a | - | n/a | 999 | | |
| BCLL 0.0 * | Rep Stress Incr | YES | WB | 0.10 | Horz(TL) | 0.00 | 6 | n/a | n/a | | |
| BCDL 10.0 | Code IBC2000/ANSI95 | | (Matrix) | | | | | | | Weight: 34 lb | |

**LUMBER**
TOP CHORD  2 X 4 SPF-S 1650F 1.5E
BOT CHORD  2 X 4 SPF-S 1650F 1.5E
OTHERS       2 X 4 SPF Stud

**BRACING**
TOP CHORD    Sheathed or 6-0-0 oc purlins.
BOT CHORD    Rigid ceiling directly applied or 10-0-0 oc bracing.

**REACTIONS** (lb/size)  1=-123/11-5-8, 7=-123/11-5-8, 2=396/11-5-8, 6=396/11-5-8, 9=166/11-5-8, 10=354/11-5-8, 8=354/11-5-8
Max Horz 1=-119(load case 6)
Max Uplift1=-164(load case 2), 7=-164(load case 3), 2=-102(load case 8), 6=-78(load case 9), 10=-117(load case 8), 8=-115(load case 9)
Max Grav 1=88(load case 7), 7=44(load case 9), 2=522(load case 2), 6=522(load case 3), 9=166(load case 1), 10=484(load case 2), 8=484(load case 3)

**FORCES** (lb) - Maximum Compression/Maximum Tension
TOP CHORD    1-2=-146/152, 2-3=-158/101, 3-4=-133/91, 4-5=-133/91, 5-6=-158/101, 6-7=-22/123
BOT CHORD    2-10=-22/74, 9-10=-22/74, 8-9=-22/74, 6-8=-22/74
WEBS           4-9=-142/2, 3-10=-418/138, 5-8=-418/138

**NOTES**
1) Wind: ASCE 7-98; 90mph; h=30ft; TCDL=5.0psf; BCDL=5.0psf; Category II; Exp C; enclosed; MWFRS gable end zone and C-C
   Exterior(2) zone; cantilever left and right exposed ; end vertical left and right exposed; Lumber DOL=1.33 plate grip DOL=1.33. This truss
   is designed for C-C for members and forces, and for MWFRS for reactions specified.
2) Truss designed for wind loads in the plane of the truss only.  For studs exposed to wind (normal to the face), see MiTek "Standard Gable
   End Detail"
3) TCLL: ASCE 7-98; Pf=45.0 psf (roof snow); Exp C; Partially Exp.
4) Unbalanced snow loads have been considered for this design.
5) This truss has been designed for a 10.0 psf bottom chord live load nonconcurrent with any other live loads per Table 1607.1 of IBC-00.
6) Gable requires continuous bottom chord bearing.
7) Gable studs spaced at 2-0-0 oc.
8) * This truss has been designed for a live load of 20.0psf on the bottom chord in all areas where a rectangle 3-6-0 tall by 1-0-0 wide will fit
   between the bottom chord and any other members.
9) Provide mechanical connection (by others) of truss to bearing plate capable of withstanding 164 lb uplift at joint 1, 164 lb uplift at joint 7,
   102 lb uplift at joint 2, 78 lb uplift at joint 6, 117 lb uplift at joint 10 and 115 lb uplift at joint 8.
10) SEE MiTek STANDARD PIGGYBACK TRUSS CONNECTION DETAIL FOR CONNECTION TO BASE TRUSS

**LOAD CASE(S)** Standard

**Figure 21–9.** T2PBGE engineering detail. *Courtesy of Robert Marini Builders, Inc.*

# Common Rafters

## Roof Construction

In a common frame roof, the rafters, ridge board, and collar ties are the structural members. They are sized and spaced to support the weight of the roof itself plus any snow or wind that can be expected. The protection from weather is provided by sheathing (or roof decking) and roofing material, **Figure 22–1.**

The size and spacing of the rafters vary depending on their length, pitch, and *load* (weight they must support) and are specified in Tables R802.5.1 through R802.5.8 of the *International Residential Code®.* If the rafters span a great distance or are spaced far apart, they must be deep to support their load. (The vertical dimension of rafters and joists is called *depth*). The size and spacing of the rafters are shown on a section view, **Figure 22–2.** The ridge board is also usually shown on a section view. The ridge board should be made of stock 2 inches deeper than the rafters. The greater depth is needed because an angled cut across the rafters is longer than a square cut across the rafters, **Figure 22–3.**

When the roof framing is complicated, a separate roof framing plan may be included, **Figure 22–4.** Roof framing plans are used only to show the general arrangement of the framing. Therefore, unless specific dimensions are included, framing plans should not be relied upon for the lengths of the framing members.

The roof frame may also include **collar beams,** or *collar ties.* Collar beams are usually made of 1-inch (nominal thickness) lumber. They are normally included on every second or third pair of rafters. This information should be shown on the roof section if collar beams are planned.

The slope of the roof may be shown on any section view that shows the rafter size. Slope is usually also shown on the building elevations.

## Roof Covering

The most common roof decking materials for houses are plywood and oriented-strand board (OSB). However, in post-and-beam construction where the extra strength is needed to span the distance between the rafters, dimensional lumber is used for the roof deck.

The material used to provide weather protection is often chosen for its architectural style, **Figure 22–5.** Some of the most common materials are asphalt shingles, wood shingles, and terra-cotta or metal tiles. The material to be used is shown on the section view of the roof and usually on the building

## Objectives

After completing this unit, you will be able to perform the following tasks:

O  Find information about roof construction on drawings.

O  Calculate the length of common rafters.

O  Calculate the length of rafter tails when overhang is given.

**Figure 22–1.** Elements of a roof.

CROSS SECTION C-C

SCALE: 3/8" = 1'-0"

**Figure 22–2.** A section view of the roof shows the size and spacing of rafters. *Courtesy of Home Planners, Inc.*

2 X 8 RIDGE BOARD

12
10

2 X 6 RAFTER

5 1/2"
7 1/4"

**Figure 22–3.** The ridge board must be made of wider lumber than the rafters to allow for the angled plumb cut.

2×10 @ 16" o c

WALL

2×10 @ 16' o.o.

WALL

BOX BM

2×10 @ 16' o c

SKYLT

2×10 @ 16" o.c.

2×12 RIDGE

WALL

WALL

5
6

2×8 RIDGE

2×10 @ 16" o.c.

2×10 @ 16" o.c.

BOX BM

CHIM

BOX BM

2×10@16" o.c.

6
6

ROOF FRAMING PLAN

**Figure 22–4.** Roof framing plans show the arrangement of rafters. *Courtesy of Robert C. Kurzon.*

length of a **common rafter**, you must know the run of the rafter and its rise per foot of run.

The run can be found on the floor plan of the building. The rise per foot of run is shown on the building elevations. The length of the common rafter is then found by following these steps:

Step 1.  Find the number of inches of rise per foot of run at the top of the table. These numbers are the regular graduations on the square.

Step 2.  Under this number, find the length of the rafter per foot of run. A space between the numbers indicates a decimal point.

Step 3.  Multiply the length of the common rafter per foot of run (the number found in Step 2) by the number of feet of run.

Step 4.  Add the length of the tail, and subtract one-half the thickness of the ridge board. The result is the length of the common rafter as measured along the measuring line.

**Note:** If the overhang is given on the working drawings, it can be added to the run of the rafter instead of adding the length of the tail.

**Example:** Find the length of a common rafter for the roof in **Figure 22–7.**

**Figure 22–5.** Tile roofs are popular in some parts of the country.

elevations. The roofing should be applied over a layer of asphalt-saturated building paper—sometimes called *slater's felt.*

Asphalt roofing materials, including felt and shingles, are sold by their weight per hundred square feet of coverage. One hundred square feet of roof is called a **square**. If enough shingles to cover one square weigh 235 pounds, they are called 235-lb shingles. The felt used under roofing is typically 15-lb or 30-lb weight.

## Finding the Length of Common Rafters

Carpenters use a rafter table to find the length of rafters. These tables are available in handbooks and are printed on the face of a framing square, **Figure 22–6.** To find the

| | 18 | 17 | 16 | 15 | 14 | 13 | 12 |
|---|---|---|---|---|---|---|---|
| LENGTH COMMON RAFTERS PER FOOT RUN | 21 63 | 20 81 | 20 | 19 21 | 18 44 | 17 69 | 16 97 |
| " HIP OR VALLEY " " | 24 74 | 24 02 | 23 32 | 22 65 | 22 | 21 38 | 20 78 |
| DIFF IN LENGTH OF JACKS 16 INCHES CENTERS | 28 7/8 | 27 3/4 | 26 11/16 | 25 5/8 | 24 9/16 | 23 9/16 | 22 5/8 |
| " " " 2 FEET | 43 1/4 | 41 5/8 | 40 | 38 7/16 | 36 7/8 | 35 3/8 | 33 15/16 |
| SIDE CUT OF JACKS USE | 6 11/18 | 6 15/16 | 7 3/16 | 7 1/2 | 7 13/16 | 8 1/8 | 8 1/2 |
| " HIP OR VALLEY | 8 1/4 | 8 1/2 | 8 3/4 | 9 1/16 | 9 3/8 | 9 5/8 | 9 7/8 |

| | 12 | 11 | 10 | 9 | 8 | 7 | 6 | 5 | 4 | 3 | 2 | 1 |
|---|---|---|---|---|---|---|---|---|---|---|---|---|
| | 16 97 | 16 28 | 15 62 | 15 | 14 42 | 13 89 | 13 42 | 13 | 12 65 | 12 37 | 12 16 | |
| | 20 78 | 20 22 | 19 70 | 19 21 | 18 76 | 18 36 | 18 | 17 69 | 17 44 | 17 23 | 17 09 | |
| | 22 5/8 | 21 11/16 | 20 13/16 | 20 | 19 1/4 | 18 1/2 | 17 7/8 | 16 7/8 | 16 1/2 | 16 1/4 | | |
| | 33 15/16 | 32 9/16 | 31 1/4 | 30 | 28 7/8 | 27 3/4 | 26 13/16 | 26 | 25 5/16 | 24 3/4 | 24 5/16 | |
| | 8 1/2 | 8 7/8 | 9 1/4 | 9 5/8 | 10 | 10 3/8 | 10 3/4 | 11 1/16 | 11 3/8 | 11 5/8 | 11 13/16 | |
| | 9 7/8 | 0 1/8 | 10 3/8 | 10 5/8 | 10 7/8 | 11 1/16 | 11 5/16 | 11 1/2 | 11 11/16 | 11 13/16 | 11 15/16 | |

**Figure 22–6.** Rafter table on the face of a square: The top line is the length of common rafters per foot of run.

Figure 22–7. Find the length of a common rafter.

1. Rise per foot of run = 4"

2. Length of common rafter per foot of run = 12.65"

3. Run of one rafter including overhang = 16'-0"

4. 16 × 12.65" = 202.40" (round off to 202½")

5. Subtract one-half the thickness of the ridge board: 202½" − ¾" = 201¾"

## Shed Roof Framing

Most of the Lake House has shed roof construction. That is, the common rafters cover the full span of the area they cover in a single slope. To find the length of shed rafters, treat the entire span of the rafters as run. For example, the total width of the garage is 13'-9". There is no overhang on either side of the garage, so the span of the garage rafters is 13'-9". Use 13'-9", or 13.75 feet, as the run in calculating the length of common rafters. (See Math Review 20.)

Where shed rafters butt against a wall or other vertical surface, the drawings should include a detail to show how they are fastened. **Figure 22–8** shows three methods of fastening rafters to a vertical surface.

## Roof Openings

It is often necessary to frame openings in the roof. Where chimneys, skylights, or other features require openings through the rafters; headers and double framing members are used. This method of framing openings is similar to that used in floor framing.

## ✔ CHECK YOUR PROGRESS

Can you perform these tasks?

- List the size and spacing of rafters, ridge boards, and collar beams.
- Describe the framing of a gable with a rake overhang.
- List the materials to be used for roof covering.
- Use a rafter table to find the length of a common rafter, including the rafter tail.

**Figure 22–8.** Details of shed roof to wall.

## ASSIGNMENT

Give the following information for the rafters of the Lake House:

| Rafter Location | Thickness × Depth | Run | Rise per Foot Run | Length | O.C. Spacing |
|---|---|---|---|---|---|
| 1. Kitchen | | | | | |
| 2. Bedroom #2 and Loft | | | | | |
| 3. Living room | | | | | |
| 4. Bedroom #1 | | | | | |
| 5. Garage | | | | | |

# Hip and Valley Framing

## Hip Rafters

**Hip rafters** run from the corner of the building to the ridge at a 45° angle, **Figure 23–1.** The length of hip rafters can be found by using a table found on most framing squares, **Figure 23–2.** The second line of this table is used to calculate the length of hip rafters. This table is based on the unit-run-and-rise method for finding the length of common rafters, explained in Unit 22.

To calculate the length of a hip rafter, you must know the run of the common rafters in the roof and the unit rise of the roof, **Figure 23–3.** The length of the hip rafters is then found by using the table for length of hip and valley rafters in the same way the table for the length of common rafters is used in Unit 22.

Step 1. Find the unit rise (number of inches of rise per foot of run) at the top of the table. These numbers are the regular graduations on the square.

Step 2. Under this number, find the length of the hip rafter per foot of run of the common rafters.

Step 3. Multiply the length of the hip rafter per foot of common-rafter run (the number found in Step 2) by the number of feet of run of the common rafters (one-half the width of the building).

Step 4. Subtract the ridge allowance. Because the hip rafter meets the ridge board at a 45° angle, the ridge allowance is one-half the 45° thickness of the ridge board, **Figure 23–4.** The 45° *thickness* is the length of a 45° line across the thickness of the ridge board. The 45° thickness of a 1½-inch (2-inch nominal) ridge board is 2⅛ inches. Therefore, the ridge allowance for a hip rafter on a 1½-inch ridge board is 1¹/₁₆-inches.

**Note:** If the hip rafter includes an overhang, add the overhang of the common rafters to the run of the common rafters.

## Objectives

After completing this unit, you will be able to perform the following tasks:

O Calculate the length of hip rafters.

O Calculate the length of valley rafters.

O Calculate the length of hip and valley jack rafters.

**Figure 23–1.** Parts of a hip and valley roof frame.

RIDGE BOARDS

COMMON RAFTERS

VALLEY JACK RAFTERS

VALLEY RAFTER

HIP RAFTERS

HIP JACK RAFTERS

**Figure 23–2.** Rafter table on the face of a square.

**Example:** Find the length of a hip rafter for the roof shown in **Figure 23–5.**

1. Rise per foot of run = 4″

2. Length of hip rafter per foot of common-rafter run = 17.44″

3. Common-rafter run including overhang = 16′-0″

4. 16 × 17.44″ = 279.04″ (round off to 279¹/₁₆″)

5. Subtract ³/₄″ ridge allowance: 279¹/₁₆″ − ³/₄″ = 278⁵/₁₆″

Figure 23–3. To use the table for the length of hip and valley rafters, use the run of the common rafters in the roof.

The rafters that butt against the hip rafter are called hip **jack rafters**; they are cut at an angle, **Figure 23–6.** This angled cut produces a surface that is longer than the width of the lumber from which the rafter is cut. Therefore, the hip rafters are sometimes made of wider lumber than the common rafters and jack rafters.

## Valley Rafters

The line where two pitched roofs meet is called a **valley**. The rafter that follows the valley is a **valley rafter.** It

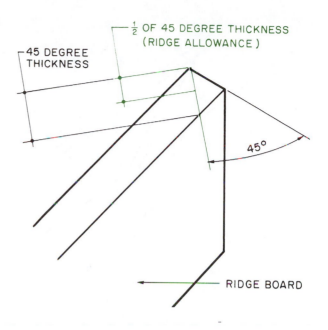

Figure 23–4. Use one-half of the 45° thickness of the ridge as a ridge allowance for hip rafters.

is most common for both roofs to have the same pitch. This results in the valley rafter being at a 45° angle with both ridges—the same angle as a hip rafter. Because the angles of the hip and valley rafters are the same and the pitch is the same, the same table can be used to compute their lengths.

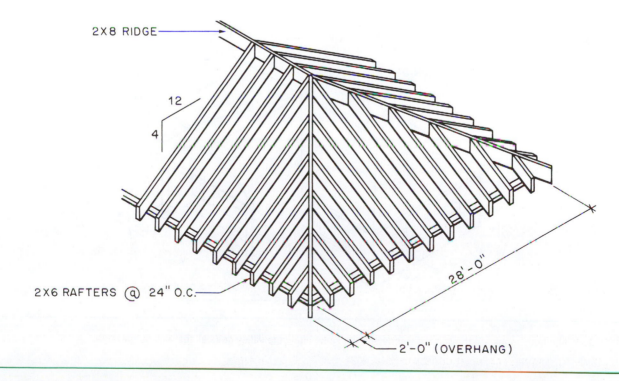

Figure 23–5. Find the length of a hip rafter.

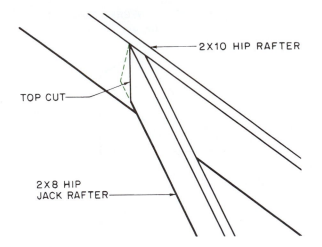

Figure 23–6. Because the top of the hip jack is cut on an angle, the hip rafter must be wider.

2X10 HIP RAFTER

TOP CUT

2X8 HIP JACK RAFTER

Figure 23–7. On some roofs the common rafter tails are close enough so no tail is needed on the valley rafters.

All steps of the procedure given earlier for hip rafters can be followed to find the length of valley rafters. However, the valley rafters often have no tail even though the roof has an overhang, **Figure 23–7**. In this case, the total length of the valley rafters is computed on the basis of the run of the common rafters excluding the overhang.

When both roofs have the same span and rise, the valley extends from the eave to the ridge, **Figure 23–8**. When one roof has a greater span than the other, the valley does not reach the ridge, **Figure 23–9**. In this case, the valley is framed in one of two ways. One valley rafter can extend to the ridge, and the other can butt against the

SAME PITCH BOTH ROOFS

VALLEY EXTENDS TO RIDGE

SPAN = x

SPAN = x

EQUAL SPANS

Figure 23–8. When both spans are the same distance and the same pitch, the valley goes all the way to the ridge.

**Figure 23–9.** When the spans are not equal, the valley does not reach the ridge.

**Figure 23–10.** Framing valleys with long and short valley rafters.

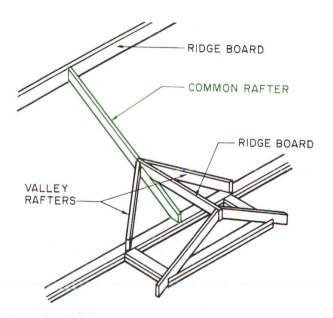

**Figure 23–11.** A valley framed against a common rafter.

first, **Figure 23–10.** The other method is to install a common rafter on the wider roof in line with the ridge of the narrower roof. Both valley rafters can then butt against this common rafter, **Figure 23–11.**

The length of the valley rafters is based on the run of the common rafters that have their upper (ridge) ends at the same level as the valley rafter. In other words,

when the long and short valley rafters are used, the length of the long valley rafter is based on the run of the common rafters in the wider roof. The length of the short valley rafter is based on the run of the common rafters in the narrower roof. When both valley rafters butt against a common rafter of a higher roof, their lengths are based on the run of the common rafters in the narrower roof.

# Jack Rafters

Rafters that extend from the wall plate to a hip rafter are called *hip jack rafters,* as shown in **Figure 23–1.** Those that extend from a valley rafter to the ridge board are called *valley jack rafters,* also shown in **Figure 23–1.**

The third and fourth lines of the rafter table on most framing squares, shown in **Figure 23–2,** are used to calculate the length of jack rafters. The length of each jack rafter in a roof varies from the length of the one next to it by the same amount, **Figure 23–12.** The amount of this variance depends on the spacing of the rafters and the pitch of the roof. The third line of the rafter table is used when the rafters are spaced 16 inches O.C.; the fourth line is used when they are spaced 24 inches O.C. As with the other lines of the rafter table, the inch numerals at the top of the square are used to indicate the unit rise of the roof. For example, if a roof has a 6 in 12 slope and the roof framing is 16 inches O.C., the difference in the length of jack rafters is 17 7/8 inches.

The first jack rafter should be a full space (16 inches or 24 inches) from the bottom of the hip or the top of the valley. Therefore, the first jack rafter should be the length shown on the rafter table.

The length of the tail must be added to the theoretical length from the table. The length of the tail on hip jacks is the same as the length of the tail on common rafters. Therefore, the length of the tail can be found by using the table for the length of common rafters. Simply treat the tail as a very short common rafter. If the overhang is not in even feet, divide the length of common rafter per foot of run on the table by 12. This gives you the length of the common rafter (or jack rafter tail) per inch of run.

**Figure 23–12.** Each jack rafter in a string varies from the next by the same amount.

**Example:** Find the length of the tails of the jack rafters in **Figure 23–13**.

1. Unit rise (rise per foot of run) = 4"

2. Length of common rafter per foot of run = 12.65"

3. Length of common rafter per inch of run = 12.65"/12 = 1.05"

4. Run of rafter tail (overhang) = 8"

5. Length of rafter tail = 8 × 1.05" = 8.4", approximately 8$\frac{7}{16}$".

Hip jack rafters are also shortened at the top to allow for the thickness of the hip rafter they butt against. This allowance is one-half the 45° thickness of the hip rafter. Valley jack rafters are shortened at the bottom to allow for the thickness of the valley rafter. This allowance is one-half the 45° thickness of the valley rafter. The valley jack rafters are also shortened at the top to allow for the thickness of the ridge board. The ridge board allowance for valley jacks is the same as the ridge board allowance for common rafters—one-half the actual thickness of the ridge board.

**Example:** Find the length of the hip jack rafters (A, B, and C) and the valley jack rafters (D, E, and F) in **Figure 23–13**.

*Hip Jack Rafters*

1. Rise per foot of run = 4"

2. Spacing of rafters = 24" O.C.

3. Difference in the length of jacks = 25$\frac{5}{16}$"

4. Theoretical length of hip jack rafter A = 0 + 25$\frac{5}{16}$" = 25$\frac{5}{16}$"

5. Add tail as found in earlier example: 25$\frac{5}{16}$" + 8$\frac{7}{16}$" = 33$\frac{3}{4}$"

6. Subtract one-half the 45° thickness of the hip rafter: 33$\frac{3}{4}$" − 1$\frac{1}{16}$" = 32$\frac{1}{16}$" (actual length of A)

7. Actual length of hip jack B = 32$\frac{11}{16}$" (length of A) + 25$\frac{5}{16}$ (from rafter table) = 58"

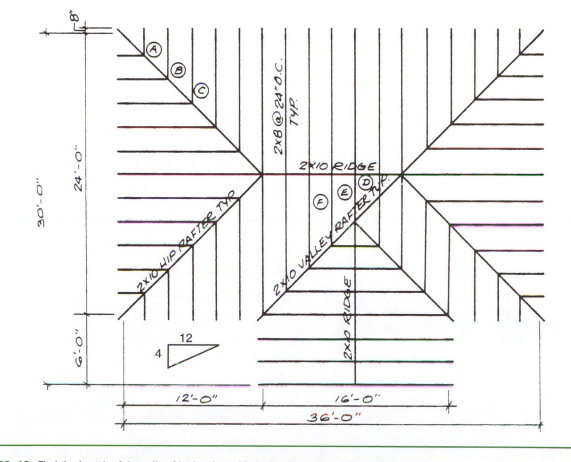

**Figure 23–13.** Find the length of the tails of jack rafters, hip jack rafters, and valley jack rafters.

8. Actual length of hip jack C = 58″ (length of B) + $25^5/_{16}″$ = $83^5/_{16}″$

*Valley Jack Rafters*

9. Theoretical length of valley jack rafter D = 0 + $25^5/_{16}″$ = $25^5/_{16}″$

10. Subtract one-half the 45° thickness of the valley rafter: $25^5/_{16}″ - 1^1/_{16}″$ = $24^1/_4″$

11. Subtract one-half the actual thickness of the ridge board: $24^1/_4″ - ^3/_4″$ = $23^1/_2″$ (actual length of D)

12. Actual length of valley jack E = $23^1/_2″$ (length of D) + $25^5/_{16}″$ (from rafter table) = $48^{13}/_{16}″$

13. Actual length of valley jack F = $48^{13}/_{16}″$ (length of E) + $25^5/_{16}″$ = $74^1/_8″$

---

## ✓ CHECK YOUR PROGRESS

Can you perform these tasks?

● Use a rafter table to find the length of hip rafters.

● Use a rafter table to find the length of valley rafters.

● Use a rafter table to find the length of hip jack rafters and valley jack rafters.

## ASSIGNMENT

A. Refer to **Figure 23–14** to complete questions 1–11.

1. What is the run of the common rafters at A?
2. How much overhang does the roof have?
3. What is the actual length of the common rafters at A?
4. What is the actual length of the hip rafter at B?

**Figure 23–14.**

NOTE: ALL SLOPE 6 ⟋ 12 TYP.

5. What is the run of the common rafters at C?
6. What is the actual length of the common rafters at C?
7. What is the length of the short valley rafter?
8. What is the actual length of the shortest hip jack rafter?
9. What is the actual length of the second shortest hip jack rafter?
10. What is the actual length of the shortest valley jack rafter?
11. What is the actual length of the second shortest valley jack rafter?

B. Refer to the Lake House drawings (in the packet) to complete this part of the assignment.
12. What is the length of the structural steel hip rafter over the dining room?

*Notes:*

o This hip rafter is a steel channel shown on details 5/6 and 6/6, and marked as MC8X8.7
o Remember to allow for the distance from the column centerline and the end of the rafter as dimensioned on the detail drawing.
o This roof has an unusual pitch of 2.96 in 12. This is close enough to use 3 in 12 for calculating rafter lengths.

# UNIT 24

# Cornices

## Objectives

After completing this unit, you will be able to perform the following tasks:

○ Describe the cornice construction shown on a set of drawings.

○ List the sizes of the individual parts of the cornice shown on a set of drawings.

○ Describe the provisions for attic or roof ventilation as shown on a set of drawings.

## Types of Cornices

The *cornice* is the construction at the place where the edge of the roof joins the sidewall of the building. On hip roofs, the cornice is similar on all four sides of the building. On gable and shed roofs, the cornice follows the pitch of the end (**rake**) rafters. The cornice on the ends of a gable or shed roof is sometimes called simply the *rake,* **Figure 24–1.** The three main types of cornice are the *box cornice,* the *open cornice,* and the *close cornice.*

## Box Cornice

The box cornice boxes the rafter tails. This type of cornice includes a fascia and soffit, **Figure 24–2.** The **fascia** covers the ends of the rafter tails. The **soffit** covers the underside of the rafter tails. There are three types of box cornices. These types vary in the way the soffit is applied.

### Sloping Box Cornice.

In the sloping box cornice, the soffit is nailed directly to the bottom edge of the rafter tails. This causes the soffit to have the same slope or pitch as the rafter, **Figure 24–3.**

### Narrow Box Cornice.

In the narrow box cornice, the rafter tails are cut level. The soffit is nailed to this level-cut surface, **Figure 24–4.**

### Wide Box Cornice.

In a wide box cornice, the overhang is too wide for a level cut on the rafter tails to hold the full width of the soffit. In conventional wood framing, *lookouts* are installed between the rafter ends and the sidewall. The lookouts provide a nailing surface for the soffit, **Figure 24–5.** For a metal soffit, special metal channels fastened to the sidewall and the back of the fascia hold the soffit, **Figure 24–6.**

## Open Cornice

In an open cornice, the underside of the rafters is left exposed, **Figure 24–7.** Blocking is installed between the rafters and above the wall plate to seal the cornice from the weather. An open cornice may or may not include a fascia.

**Figure 24–1.** The cornice is the construction at the place where the roof and the sidewall meet.

**Figure 24–2.** Parts of a box cornice.

**Figure 24–3.** Sloping box cornice.

## Close Cornice

In a close cornice, the rafters do not overhang beyond the sidewall, **Figure 24–8.** The interior may be sealed by the sheathing and siding or by a fascia. In either case, there must be some provision for ventilation.

## Cornice Returns

Any type of construction described for cornices can be used for the rake. When a sloping cornice is used, the fascia and soffit follow the line of the roof up the rake. When a level box cornice is used, a **cornice return** is necessary.

**Figure 24–4.** Narrow box cornice.

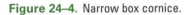

**Figure 24–5.** Wide box cornice with horizontal lookouts.

This is the construction that joins the level soffit and fascia of the eave with the sloping rake, **Figure 24–9.**

The style of cornice return is shown on the building elevations, **Figure 24–10.** Although good architectural drafting practice requires details of all special

construction, many architects do not include details of cornice returns. The carpenter is expected to know how to achieve the desired results. **Figures 24–11** and **24–12** show the construction of two popular types of cornice returns.

**Figure 24–6.** Metal and vinyl soffit systems include ventilated soffits and channels.

**Figure 24–8.** Close cornice.

**Figure 24–7.** Open cornice.

# Ventilation

The cornice usually allows for ventilation of the attic or roof. Attic or roof ventilation is necessary in both hot and cold weather. Without ventilation, the air in the attic becomes stagnant because it is trapped and unable to circulate.

In hot weather, this stagnant air builds up heat and makes the house warmer. Hot, stagnant air can hold a large amount of moisture. When this moisture-laden air

**Figure 24–9.** Cornice return. *Courtesy of Richard T. Kreh, Sr.*

16 × 24" SCREENED LOUVERED VENT

1×10 HEAD BD.
3/4" HALF-ROUND
1×4

36/30 D.H.

20" SHUTTERS

1×12 BOARD SIDING
1×3 BATTENS

**Figure 24–10.** The cornice returns can be seen on the building elevations.

AL. FLASHING
COMP. SHINGLES
3/4" PLYWD.
3/4" X 3 1/4" CROWN MOLD
1×6
2×4 BLOCKING
1/2" PLYWD.
3/4" X 2 1/4" COVE MOLD
1×8
1/2" SIDING
1/2" SHEATHING

2X4 @ 16" O.C.
2-2X4
3 1/2" BATT INSUL.
1/2" GYP. BD.
2X4 @ 16" O.C.

**Figure 24–11.** Section through cornice return shown in Figure 24–10.

comes in contact with the cooler roof, the moisture condenses. The condensation can reduce the effectiveness of the insulation. Condensation can also cause the wood in the attic to rot.

Attic or roof ventilation also helps prevent ice buildup in cold climates. Without ventilation, the heat from the building melts the snow that falls on the roof. As the melted snow reaches the overhang of the roof, it refreezes. Eventually an ice dam may build up. The ice dam can back up newly melted snow, causing it to seep under the shingles, **Figure 24–13.**

Ventilation is created by allowing cool air to enter through the cornice and exit at the ridge or through special ventilators, **Figure 24–14.** A section view of

Figure 24–12. Framing for cornice return shown in Figure 24–9.

RAKE RAFTER

2X4 BLOCKING
(NAIL TO CORNER STUD)

2X4 LOOKOUT

WARM AIR OUT

WARM, HUMID
ATTIC AIR

COOL
AIR IN

Figure 24–14. Airflow through the attic.

MELTED SNOW

WATER BACKS UP

ICE DAM

WARM
ATTIC

HEATED SPACE

Figure 24–13. Ice dam.

the sidewalls or a special roof and cornice detail shows the construction of the cornice including any ventilators. Notice that each of the soffits in **Figures 24–2, 24–4, and 24–5** includes a ventilator.

The heated air can be allowed to exit in one of three ways. Some buildings have a ventilated, metal ridge cap. This may be shown on the building elevations or on a separate detail. When a ventilated ridge is used, an opening is left in the roof decking at the ridge. Metal roof ventilators can be installed in the roof. Ventilators can be installed in the gables. Metal roof ventilators and gable ventilators are usually shown on the building elevations only.

## ✓ CHECK YOUR PROGRESS

Can you perform these tasks?

- Describe the construction of a typical sloping box cornice.
- Describe the construction of a typical narrow box cornice.
- Describe the construction of a typical wide box cornice.
- Describe the construction of a typical open cornice.
- Describe the construction of a typical close cornice.
- Describe the construction of a return for each of the above cornice types.
- Trace the flow of ventilating air from the point it enters a cornice to the point it exits the attic.

## ASSIGNMENT

Refer to the Two-Unit Apartment and the Lake House drawings (in the packet) to complete the assignment.

1. Which type of cornice does the Apartment have?
2. What material is used for the Apartment cornice?

3. How wide is the Apartment soffit?

4. The Apartment fascia is made of two parts. What are they?

5. What provision does the Apartment cornice have for ventilation?

6. How does attic air exit from the Apartment?

7. Sketch the Lake House cornice, and show where air enters for ventilation.

8. There are two ways that air can escape from the Lake House roof. Describe one.

# Windows and Doors

## Window Construction

Most windows are supplied by manufacturers as a completely assembled unit. However, the carpenters who install windows often have to refer to window details for information. Some special installations require knowledge of the construction of the window unit. Also, when a special window is required, the carpenter may build parts of it on the construction site.

## Wood Windows

The major types of windows are briefly discussed in Unit 19. All these windows include a frame and sash. The **sash** is the glass and the wood (or metal) that holds the glass. The sash is made of **rails** (horizontal parts) and **stiles** (vertical parts), **Figure 25–1.** The sash may also include muntins. **Muntins** are small

## Objectives

After completing this unit, you will be able to perform the following tasks:

○ Interpret information shown on window and door details.

○ Find information in window and door manufacturers' catalogs.

**Figure 25–1.** Parts of a window

strips that divide the glass into smaller panes. The glass is sometimes called the *lite.*

The window frame is made of the *side jambs,* the *head jamb,* and the **sill.** Stop molding is applied to the inside of the jambs to hold the sash in place. Factory-built windows also come with the exterior casing installed. The **casing** is the molding that goes against the wall around the frame. The interior casing and the apron, if one is included, are applied after the window is installed.

## Metal Windows

Many buildings have vinyl or metal windows. Improvements in the design of metal windows have made them competitive with wood windows in both cost and energy efficiency. The most important of these design improvements has been the development of thermal-break windows. **Thermal-break windows** use a combination of air spaces and materials that do not conduct heat easily to separate the exterior from the interior, **Figure 25–2.**

The basic parts of a metal or vinyl window are similar to those of a wood window. The sash consists of a stile, rails, and glazing. The frame is made up of side jambs, a head jamb, and a sill. However, the trim (casing) is not included as part of the window. Often the window frame itself is the only trim used on the exterior. The frame includes a nailing fin for attaching the window to the building framing.

## Window Details

All windows include the parts discussed so far. However, to show the smaller parts, which vary from one window style to another, architects and manufacturers use detail drawings. The most common type of window detail is a section, **Figure 25–3.** All the parts can be shown in section views of the head, sill, and one side jamb. These sections also usually show the wall framing around the window.

Figure 25–2. Thermal-break windows use insulating materials and air spaces to separate the interior from the exterior. *Courtesy of Ethyl Capitol Products Corporation.*

**Figure 25–3.** Typical window detail drawing. *Courtesy of Andersen Corporation, Bayport, MN 55003.*

Some of the parts that can be found on window sections are defined here. Find each of the parts on the sections and illustrations in **Figures 25–4, 25–5,** and **25–6.**

○ *Weather stripping* is used on windows that open and close. It forms a weather-tight seal around the sash.

○ The *transom bar* is the horizontal part of a window frame that separates the upper and lower sash.

○ *Meetings rails* or check rails are the rails that meet in the middle of a **double-hung window.**

○ *Insulating glass* is a double layer of glass, creating a dead-air space. The dead air acts as an insulator. Better windows are filled with inert gas, such as argon, to provide better insulating qualities.

○ *Extension jambs* are fastened to standard jambs when the window is installed in a thicker than normal wall.

○ A **mullion** is a vertical section of the frame that separates side-by-side sash. If the mullion is formed by butting two windows together, it is called a narrow mullion. If the mullion is built around a stud or other structural support, it is called a support mullion.

**Figure 25–4.** Window parts.

Details for metal windows are often drawn as simplified sections of the frame only. **Figure 25–7** shows typical details for the thermal-break window in **Figure 25–2.**

AWNING DETAILS

scale: 3" = 1'-0"

Courtesy of Andersen Corporation, Bayport, Minnesota 55003

**Figure 25–5.** *Courtesy of Andersen Corporation, Bayport, MN 55003.*

# Door Construction

Doors include many of the same basic parts as windows, **Figure 25–8**. A door frame consists of side jambs and a head jamb with **stop** and casing. Exterior door frames also include a sill. Many doors are made of a framework with panels, **Figure 25–9**. The parts of **panel doors** are named similarly to the parts of a window. The vertical parts are *stiles,* and the horizontal parts are *rails.* Doors with glass or louvers are variations of panel doors. The framework is made of rails and stiles, and the glass or louvers replace the panels. Several manufacturers make molded doors. The most common type of molded door is made of hardboard for interior doors or steel for exterior doors which is manufactured in folds that contour the surface to look like panel doors, **Figure 25–10**. *Hollow core doors* consist of an internal frame with flat "skin" applied to each side, **Figure 25–11**. Some exterior doors have insulation between the two outer steel skins. These insulated doors result in considerable heating and cooling savings.

# Door Details

Door details are usually less complex than window details. Where security, fire alarm, electronic lock, or special systems must be run in door and window metal frames, the corresponding installers must familiarize themselves with the actual manufacturer details and coordinate their installation with the door and window installation schedule. Carpenters rarely make doors, so all that is needed are simple details of the door frame and its trim, **Figure 25–12**.

Many doors are sold as prehung units. In these units the frame is assembled, including the trim, and the door is hung in the frame. A section view of the jambs shows

Vinyl Flashing Applied

3/8"

HEAD

Groove Glazed Pat. No's. U.S. 3,556,542 Can. 903,010 Other Patents Applied For.

UNITS DIMENSION HEIGHT

ROUGH OPENING HEIGHT

CHECK RAIL

Weather Stripping

Stool Optional

Vinyl Sill Windbreak Applied

1/4"

Apply Caulking

SILL

**INSTALLATION DETAILS**
scale: 3" = 1'-0"

**Figure 25–6.** *Courtesy of Andersen Corporation, Bayport, MN 55003.*

# Reading Catalogs

It is often necessary to find specific information about windows or doors in the manufacturer's catalog. Usually the catalog has a table of contents listing the types of windows and doors shown. **Figure 25–13** shows typical pages reprinted from a manufacturer's catalog. A careful reading of these sample pages will help you to find the information you need in manufacturers' catalogs. For each type of window or door, you will find some or all of the listed information:

○ A brief description of the window type and some of the features the manufacturer wants to highlight—a little advertising.

○ Installation detail drawings.

○ Sizes available—This information usually consists of drawings of the various sizes and arrangements, with dimensions for glass size, stud or rough opening, and unit dimensions. (Notice that the nominal size in the catalog sample is written as a four-digit number. The first two digits are the approximate width in feet and inches. The last two digits are the approximate height in feet and inches. For example, a 6030 window is roughly 6 feet wide by 3 feet high.)

○ Additional information, such as optional equipment available.

how the door is installed. For example, the door detailed in **Figure 25–12** is made with two-piece jambs. These split jambs are pulled apart; each side is then slid into the opening for installation. The stop can be either applied or integral. *Applied stop* is molding that is applied to the jambs with finish nails. *Integral stop* is milled as a part of the jamb when the jamb is manufactured.

**Figure 25–7.** Typical metal window details. *Courtesy of Ethyl Capitol Products Corporation.*

**Figure 25–8.** Parts of a door frame.

**Figure 25–9.** Construction of a panel door.

Figure 25-10. Molded hardboard door.

2 1/4" RANCH CASING
6 1/2" SPLIT JAMB
1 1/8" RANCH STOP
1 3/4" DOOR
WEATHERSTRIP
2-2X6
BRICK MOLD

**EXT. DOOR**
SCALE: 3" = 1'-0"

**Figure 25-12.** Typical jamb detail for an exterior door.

SOLID FRAME
HONEYCOMB PAPER CORE
(INSULATED FOR
EXTERIOR DOORS)
SKIN
HOLLOW CORE

SOLID FRAME
SOLID WOOD BLOCKS
SKIN
SOLID CORE

**Figure 25-11.** Flush doors.

**THERMAL-BREAK FRAME SECTIONS**—Outside frames at the nailing fins are separated by rigid foamed poly-vinyl chloride, roll pressed to the metal frames. Inside, the frames are separated with extruded poly-vinyl chloride and weather stripped with mylar shielded pile. "Push-pull", panel retaining pin is shown in its position in head section. Pin allows easy removal, and locking into position of the fixed panel (no tools are needed). Integral nailing fins surround the entire frame for easy window installation.

# CAPITOL'S
# *E·700 ALUMINUM
# INSULATED GLASS, THERMAL-BREAK, ROLLING WINDOW

Rated HS-B2-HP (residential)   Rated HS-A2-HP (architectural)

The latest addition to Capitol's fine line of aluminum thermal-break products, the E-700 affords the benefits of thermal-break performance and the convenience of insulated glass . . . frost-free window frames . . . virtual elimination of window glass condensation. In addition to its excellent "energy saving" qualities, the E-700 offers these "sales appealing" features: Window can install with or without sash panels mounted—panels can be installed later from inside the building . . . Panels can be removed and replaced without tools and fixed panel is held in place with a finger operated "push-pull" retaining pin . . . Roll panel latches and automatically locks in closed position. Windows are available in arrangements such as side vents, center vents, sub-, side-, transom lites and fixed/removable lites. All models are available in PPG enamelized and baked-on colors (standard colors at no added cost).

*U.S. Patent No. 4,151,682

HORIZONTAL OR VERTICAL WINDOW MULLING

contact Capitol for complete details.

STACK MULL

SIDE BY SIDE MULL

5" PARTITION MULL

**POSITIVE ACTION, SELF-LOCKING LATCH**—Finger-tip ease of operation with this attractive, rugged lock-latch. Latch is located on the interlock stile of the rolling panel and is constructed of Celcon, Ultra-violet stabilized, acetal copolymer M90-08. Latch automatically locks the window when panel is moved into closed position.

Exterior screen of fiberglas mesh, can be inserted or removed from inside the building without removing glass panels. (Optional at added cost)

**INTERLOCKING MEETING STILES**—The key to the thermal-break feature, at the panel interlocks, is revealed in this photo. Fixed sash interlock stile is at left and rolling sash interlock stile at the right. The stile interlocks are insulated and parted with rigid Celcon separator to maintain thermal-break performance. Insulated glass is cushioned in flexible PVC and full length flexible vinyl flaps seal panels at each interlock.

**Figure 25–13.** Pages are reprinted from a typical manufacturer's catalog. *Courtesy of Ethyl Capitol Products Corporation.*

# Fully Effective Thermal Performance Without Breaking (weakening) the Fixed and Rolling Panels!

This detail shows, that we have, without exception, succeeded at each and every part of this design to completely separate the exterior "cold" window surfaces from the interior "warm" window surfaces - both at frame, sash and glass, by the use of enclosed air spaces (as the most effective insulation) and by applied low conductive plastic insulators. And this has been achieved by only thermo-breaking the frame around the center divider and by shielding the unbroken sashes with the divider either from their exposure to the inside or to the outside of the building.

In other words: The fixed outside (cold) sash has been insulated from the interior, and the rolling (warm) sash has been insulated from the exterior.

And we can confirm our theories with a thermal test report by the AAMA authorized Electrical Test Laboratories in New York.

□ *COLOR LINE DEPICTS THE UNIQUE METHOD THAT SEPARATION IS MAINTAINED BETWEEN INSIDE AND OUTSIDE TEMPERATURES.*

▨ *SHOWS WHERE LOW CONDUCTIVE INSULATING MATERIALS COME INTO USE.*

*Window rates a Condensation Resistance Factor of 50, amply exceeding the minimum 35 CRF needed to pass the Thermal Performance Specification.*

Products meet the requirements of HUD Minimum Property Standards (MPS) and the USDA, FmHA requirements.

**SILL AND ROLLING PANEL SECTION**—Sloping sill drains water to the outside and is machined to receive one-way weep valves that prevent air infiltration. The thermal-break feature between the inside and outside frame and the weatherstripping are clearly shown here. The roll panel stile shows the Celcon corner section that acts both as a housing for the roller and a guide for the panel as it moves in the sill cavity.

**ALUMINUM ROLLERS**—Celcon housing is designed to make possible a two position roller height. Roller wheels are of aluminum and ride on a rigid vinyl track for smooth panel operation. Notice that panel frames do not butt at corners. They telescope to achieve a rigid corner and eliminate gaps at the corner joints.

**FIXED PANEL REMOVAL**—Retaining pins at the frame head and sill lock the fixed window panels in place. Fixed panel cannot be removed when roll panel is in closed position. Retaining pins can be moved without the use of tools to easily remove the fixed panel.

To provide a weather tight seal at the machined areas at the top and bottom of the interlocks, "finger-operated" PVC sealer blocks are used.

**Figure 25–13.** (Continued)

## SPECIFICATIONS

**GENERAL:** Windows shown are identified as series E-700, rated HS-B2-HP (residential) and series E-705, rated HS-A2-HP (Architectural), as manufactured by Capitol Products Corporation. All horizontally rolling windows and window arrangements (side-vents, center-vents, sub-, side-, transom lites and fixed removable lites) shall be thermally insulated aluminum windows. AAMA requires that the largest size of any particular model we offer for sale be tested. That size has been tested and meets AAMA's specifications. No other size of the model will be tested and no representation as to air infiltration is made except as to the model tested. Windows furnished for job will duplicate the test window in all mechanical parts and details within standard commercial tolerances as required under ANSI/AAMA 302.9-77.

**OPERATION:** Window frames can be installed without sashes. Fixed and rolling sashes shall be loaded into frames from the building interior and be removable without the use of any tools for cleaning, maintenance or re-glazing. The rolling sash shall automatically lock when in closed position. Insect screens shall be installed or removed from the building interior without removing a sash or the use of tools, and shall be reversible for handed rolling sash operation.

**MATERIALS:** All frame members, rails and stiles shall be fabricated from extruded aluminum alloy 6063-T5 with nominal wall thicknesses of .062" for the Series E700 windows, and with nominal wall thicknesses of .062" for all members except the sill, which shall have a nominal wall thickness of .078", for the Series E705 windows.

Thermal-break frame separators shall be of extruded rigid and rigid-foamed poly-vinyl chloride. Glazing gaskets shall be 70 durometer extruded poly-vinyl chloride. Weatherstripping shall be 50 durometer extruded poly-vinyl chloride.

**WEATHERSTRIPPING:** Weatherstripping at the frame perimeters and the interlocks shall be extruded 50 durometer poly-vinyl chloride. Adjustable, dual-durometer PVC end seals shall be provided at all interlock ends at the point of contact on the frame separators.

Weather stripping at head and sill and at rolling sash to be mylar shielded pile.

Fasteners shall be stainless steel. Hardware shall be aluminum and Celcon, Ultra-violet stabilized, Acetal Copolymer M90-08. Sash bumpers shall be adhesive backed poly-vinyl chloride.

**CONSTRUCTION:** Frame members shall be thermally insulated by a crimped insulator and a snap-on type dual-durometer separator which functions as insulator and weatherstrip for the exterior fixed sash and for the rolling sash. Jambs shall be machined at both ends to receive telescopingly the head and sill. The joint shall be made watertight by applying a die-cut gasket and fastening with two screws per joint. Sills shall be machined to receive one-way valves and fixed sash retaining pins. Window thermal design to be such that no thermal break of sash aluminum components is required. All sash stiles shall be tubular aluminum shapes for torsional rigidity and shall be machined at both ends to telescope into top and bottom rails and fastened with one screw per joint together with top rail guides, fixed sash bottom rail setting block and rolling sash roller housing.

Top and bottom rails shall be machined at both ends to receive stiles and shall have weep slots for draining the bottom rail glazing cavity. Interlock stiles shall be designed to provide a pocket for the insect screen frame and shall be machined to receive the spring loaded lock and lock keeper.

**HARDWARE:** Exterior fixed sashes shall have injection molded rail guides and sash lock keeper. Interior rolling sashes shall have injection molded two-position roller housings, aluminum wheels and injection molded, spring loaded lock at interlock stile. The exterior fixed sash shall be locked in place by a push-pull retaining pin at head and sill. Sills shall have PVC one-way weep valves. Jambs shall have adhesive backed bumpers at mid-height.

**FINISH:** All aluminum frame and sash members shall have a finish that provides a smooth uniform appearance. Electrostatically applied paint over alodine base provided in standard colors at no added cost. Extruded PVC and exposed injection molded parts shall be of charcoal grey color.

**GLASS:** Glass shall be ½" sealed double glazing with standard glass thickness as required by AAMA.

**SCREENS:** Frames shall be fabricated of a roll formed aluminum rail with corners accurately mitred and fitted to a hairline joint. Screen cloth shall be fiberglas. Frames shall have 2 springloaded plungers on one side rail and a finger pull on the opposite side rail. Screens and finish on screen frames shall be available at additional cost.

**INSTALLATION:** (by others) Frames should be installed straight plumb and level without springing or twisting, and securely fastened in place in accordance with recommendations or details. Mastic or caulking compounds must be applied, before installation, between fin and adjacent construction to provide weather-tight installation and to maintain the integrity of the energy saving features of the product. Loading of sashes and final adjustments shall be made by the installers to assure proper sash operation and window performance.

| PERFORMANCE LEVEL SPECIFICATIONS | | Air Infiltration 25mph wind | | Water Resistance | | Operating Force | | Uniform Load Deflection | | AAMA requires that the largest size of any particular model we offer for sale be tested. That size has been tested and meets AAMA's specifications. No other size of the model will be tested and no representation as to air infiltration is made except as to the model tested. |
|---|---|---|---|---|---|---|---|---|---|---|
| | | measured | allowed | measured | allowed | measured | allowed | measured | allowed | |
| | | cfm/ft | | psf no flow | | pounds | | inches | | |
| | B2 | .37 | .375 | 2.86 | 2.86 | 10 | 15 | .197 | .331 | |
| | A2 | .37 | .375 | 4.34 | 3.34 | 20 | 20 | .200 | .351 | |

## TWO LITE SLIDER

## SIZE CHART
### E-700 ROLLING WINDOWS
(interior dimensions shown in inches)

### SIDE VENT

### CENTER VENT

**GLASS SIZES (in inches)**

| A | 10" x 20" | R | 34" x 32" |
| B | 10" x 32" | S | 34" x 44" |
| C | 16" x 20" | T | 34" x 56" |
| D | 16" x 32" | | |
| E | 16" x 44" | AC | 36" x 44" |
| G | 22" x 20" | AD | 36" x 56" |
| H | 22" x 32" | AF | 48" x 44" |
| J | 22" x 44" | AG | 48" x 56" |
| K | 22" x 56" | BC | 26" x 44" |
| M | 28" x 32" | BD | 26" x 56" |
| N | 28" x 44" | BF | 30" x 44" |
| O | 28" x 56" | BG | 30" x 56" |
| | | BN | 32" x 44" |
| | | BO | 32" x 56" |

**SCREEN SIZES (in inches)**

| WIDTHS | HEIGHTS |
|---|---|
| for 2-Lite | for All Styles |
| 20 = 10" | 20 = 20¾" |
| 30 = 16" | 30 = 32¾" |
| 40 = 22" | 40 = 44¾" |
| 50 = 28" | 50 = 56¾" |
| 60 = 34" | |
| for Side Vent | |
| 70 = 22" | |
| 80 = 22" | |
| 90 = 28" | |
| for Center Vent | |
| 70 = 27⁵⁄₁₆" | |
| 80 = 31⁵⁄₁₆" | |
| 90 = 35⁵⁄₁₆" | |

**Ethyl** Capitol Products

CAPITOL PRODUCTS CORPORATION

Mechanicsburg, Pennsylvania 17055 (717) 766-7661

NATIONAL SASH & DOOR JOBBERS ASSOCIATION ASSOCIATE MEMBER

CPN95558 7/80 ©

**Figure 25–13.** (Continued)

USE THIS INFORMATION TO ARRIVE AT YOUR ROUGH OPENINGS!

SCALE — ½ ACTUAL SIZE

SINGLE UNITS . . . Operating windows.
for BUTT Type installation using ½" Dry Wall.
WINDOW DIMENSION + ½" for Width + ⅜" for Height = R.O.

VERTICAL DETAIL

HEAD

HORIZONTAL DETAILS

2 PANEL-Single Slide

TYPICAL FOR A-2 SASH

ROLL PANEL

FIXED PANEL

SCREEN

OUTSIDE

JAMB

JAMB

OUTSIDE

WINDOW DIMENSION

SILL

CENTER VENT

ROLL PANEL

FIXED PANEL

SCREEN

FIXED PANEL

OUTSIDE

JAMB

JAMB

SIDE VENT

ROLL PANEL

ROLL PANEL

FIXED PANEL

SCREEN

SCREEN

OUTSIDE

JAMB

JAMB

**Figure 25–13.** (Continued)

## TYPICAL BUTT TYPE INSTALLATION DETAILS

SCALE — 3" = 1'

IMPORTANT—Mastic or caulking compounds must be applied before installation, between window fin and adjacent construction to provide weather-tight installation and maintain the integrity of the products energy saving features.

## THE CAPITOL MULLING AND FIXED LITE WINDOW WALL SYSTEM

Fin and finless windows can be combined in fixed lite and operating window configurations. Either Side-by-Side or Stack Mulled. Fixed Lites can be mixed and matched with side vent, center vent and two-lite rolling windows to achieve window wall arrangements.

Windows can be ordered with fins already removed from "mulled side" of window, or fins can easily be removed in the field for mulling.

Fixed Lite System features Removable Inserts. Glass is enclosed in its own sash frame which can be removed from the main window frames.

**Figure 25–13.** (Continued)

## ✓ CHECK YOUR PROGRESS

Can you perform these tasks?

● Identify the following parts of a window: sash, rails, stiles, muntins, glazing, side jambs, head jambs, sill, stop, exterior casing, and interior casing.

● Identify each part of a window on a window detail.

● Explain the basic features of the following door types: panel, flush, hollow core, molded, and solid core.

● Explain what is meant by applied stop and integral stop.

● Find the rough opening, unit size, and glass size for windows in a manufacturer's catalog.

## ASSIGNMENT

Refer to the Lake House drawings when necessary to complete the assignment.

1. Name the lettered parts (a through f) in **Figure 25–14.**

**Figure 25–14.** *Courtesy of Andersen Corporation, Bayport, MN 55003.*

2. What is the nominal size of the window in the south end of the Lake House dining room?

3. In the catalog sample shown in **Figure 25–13,** what is the width and height of the rough opening for a #4030 window?

4. In the catalog sample, what are the rough opening dimensions for a window with a 28-inch by 44-inch glass size?

5. In the catalog sample in **Figure 25–13,** what is the glass size of the window in bedroom #2 of the Lake House?

6. In the catalog sample, what is the R.O. for the window in bedroom #2 of the Lake House?

7. Is the exterior door in the Lake House kitchen to be prehung or site hung?

8. What type and size are the doors in the Lake House playroom closet?

# Exterior Wall Coverings

## Objectives

After completing this unit, you will be able to perform the following tasks:

○ Describe the exterior wall covering planned for all parts of a building.

○ Explain how flashing, drip caps, and other devices are used to shed water.

○ Describe the treatment to be used at corners and edges of the exterior wall covering.

## Wood Siding

Wood is a popular siding material because it is available in a variety of patterns, it is easy to work with, and it is durable. Wood siding includes horizontal boards, vertical boards, shingles, and plywood.

Boards can be cut into a variety of shapes for use as horizontal siding, **Figure 26–1.** These boards are nailed to the wall surface starting at the bottom and working toward the top. With wood siding, a starting strip of wood furring is nailed to the bottom of the wall. This starting strip holds the bottom edge of the first piece of siding away from the wall. Each board covers the top edge of the one below. The amount of each board left exposed to the weather is called the *exposure* of the siding, **Figure 26–2.**

Vertical siding patterns can be created by boards or plywood, **Figure 26–3.** These materials are applied directly to the wall with no special starting strips. However, where horizontal joints are necessary, they should be lapped with rabbet joints, or Z flashing should be applied. The flashing can be concealed with **battens, Figure 26–4.**

*Shingles* take longer to apply but make an excellent siding material. In place of a starting strip of furring, the bottom **course** (row) of shingles is doubled.

BOARD          BEVELED          LOG CABIN          MOLDED HARDBOARD

**Figure 26–1.** Horizontal siding.

Figure 26–2. The exposure is the amount of the sliding exposed to the weather.

## Fiber Cement Siding

Fiber cement siding is made of portland cement, ground sand, cellulose fiber, and additives mixed with water and formed into siding. Fiber cement siding will not rot, and it is termite proof.

Fiber cement siding looks like wood siding, **Figure 26–5.** It comes in a smooth finish, a wood grain texture, and even a rough-sawn look, and it is available in widths of 6, 7½, 9½, and 12 inches. Fiber cement siding is sometimes referred to as Hardiplank, a very popular brand of fiber cement siding. Hardiplank comes in standard 12-foot lengths, while other siding materials come in 16-foot lengths. The 4-foot difference is because fiber cement siding is much heavier than wood siding, and

BOARD & BATTEN          REVERSE BOARD & BATTEN          TEXTURE 1-11 PLYWOOD

Figure 26–3. Three popular vertical siding patterns.

Figure 26–4. Horizontal hardboard siding should be covered with battens.

PLYWOOD SIDING

BOARD OR MOLDING (BATTEN) NAILED THRU SIDING INTO SOLID FRAMING

ALUMINUM FLASHING

PLYWOOD SIDING

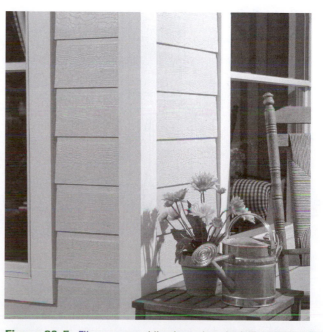

Figure 26–5. Fiber cement siding is made to look like wood siding but is more durable. *Courtesy of James Hardie Building Products.*

anything over 12 feet would be difficult for one person to handle. Although fiber cement siding looks like wood and it is installed in much the same way as wood, it is important to follow the manufacturer's installation instructions to maintain the warranty on the siding.

## Drip Caps and Flashing

It is important to prevent water from getting behind the siding, where it can cause dry rot and attract insects. Where a horizontal surface meets the siding, water is apt to collect. **Flashing** is used to prevent this water from running behind the siding. Aluminum, galvanized steel, and copper are the most common flashing materials. The flashing is nailed to the wall before the siding is applied. The lower edge of the flashing extends over the horizontal surface far enough to prevent the water from running behind the siding. Areas to be flashed are noted on building elevations. The flashing is shown on the detail drawings or building elevations, **Figure 26–6.**

The heads of windows and doors may form a small horizontal surface. Wood **drip cap** molding can be used

**Figure 26–6.** Flashing is shown on elevations and details.

in these places to shed water, **Figure 26–7.** Drip caps are shown on details and elevations.

## Corner and Edge Treatment

Regardless of the kind of siding used, the edges must be covered to prevent water from soaking into the end grain or running behind the siding. Around windows, vents, doors, and other wall openings, the trim around the opening usually covers the end grain of the siding. In an inside corner, a strip of wood can be used to form a corner bead for the siding. Outside corners in plywood, hardboard

**Figure 26–7.** Drip cap.

**Figure 26–8.** Corner boards can be used with most types of siding.

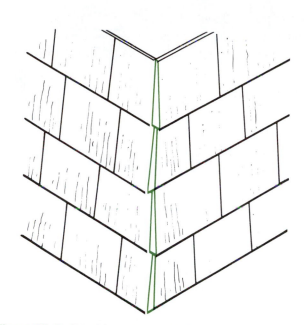

**Figure 26–9.** Shingles are trimmed to form their own corners.

sheets, or vertical boards are usually built with corner boards, **Figure 26–8.** Horizontal siding is butted against the corner boards. Shingles are trimmed to form their own outside corner, **Figure 26–9,** or they are butted against corner boards.

The corner treatment to be used can usually be seen on the building elevations. If no corner treatment is shown on the elevations, look for a special detail of a typical corner.

## Metal and Plastic Siding

Several manufacturers produce aluminum, steel, and vinyl siding and trim. The most common type is made to look like horizontal beveled wood siding. A variety of trim pieces are available for any type of application, **Figure 26–10.**

## Stucco

*Stucco* is a plaster made with portland cement. The wall sheathing is covered with waterproof building paper. Next the *lath* (usually wire netting) is stapled to the wall, **Figure 26–11.** Finally, the stucco is troweled on—a rough *scratch coat* first—**Figure 26–12,** then a *brown*

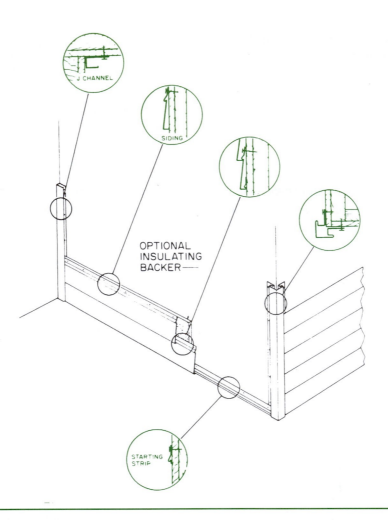

**Figure 26–10.** Accessories for metal and vinyl siding.

**Figure 26–11.** Wall prepared for stucco.

**Figure 26–12.** Scratch coat applied.

*coat* to build up to the approximate thickness, and finally a finish coat. Outside corners and edges are formed with galvanized metal beads nailed to the wall before the scratch coat is applied.

## Masonry Veneer

Masonry veneer is usually either brick or natural stone. It is called veneer because it is a thin layer of masonry over some kind of structural wall. The structure may be wood framing, metal framing, or concrete blocks. The masonry has to rest on a solid foundation. Normally the building foundation is built with a 4-inch ledge to support the masonry veneer. This ledge can be seen in a typical wall section, **Figure 26–13.**

Interesting patterns can be created by using half bricks and bricks in different positions. Some of the most common *bond patterns,* as they are called, are shown in **Figure 26–14.** If no pattern is indicated, the bricks are normally laid in running bond.

**Figure 26–13.** Typical foundation section with ledge for masonry veneer.

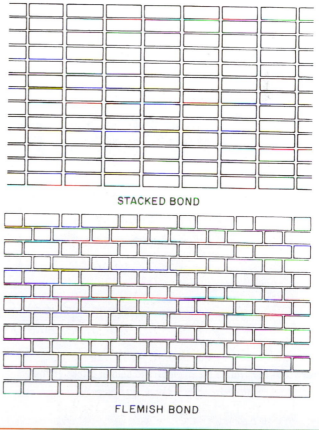

**Figure 26–14.** Frequently used bond patterns.

Above and below windows, above doors, and in other special areas, bricks may be laid in varying positions, **Figure 26–15.** If bricks are to be laid in any but the stretcher position, they will be shown on the detail drawings, **Figure 26–16.** The details for openings in masonry walls will also indicate a lintel at the top of the opening. The **lintel** is usually angle iron. It carries the weight of the masonry above the opening.

## General

The electrical installer needs to be familiar with the rough-in requirements for all outside lighting and devices and any special requirements that may be needed with

the specified exterior wall finish. The required backboard or rough-in frame and box must be properly installed to prevent any finished wall damage.

**Figure 26–15.** Brick positions.

STRETCHER    HEADER    SOLDIER    ROWLOCK

SOLDIER BRICK

3½" X 3½" STEEL LINTEL

CAULK

2X4 @ 16" O.C.

2-2X6

RANCH CASING

AWNING WINDOW (SEE SCHEDULE)

ROWLOCK BRICK

¾" INSUL. SHEATHING

1" AIR SPACE

BRICK VENEER

½" GYP. BD.

2X4 @ 16" O.C.

3½" BATT. INSUL.

TYP. WINDOW DETAIL
SCALE: 1" = 1'-0"

**Figure 26–16.** Notice the rowlock and soldier bricks on the window detail.

Can you perform these tasks?

- Describe the exterior wall covering on all parts of a building.

- List the locations of flashing on all parts of a building.

- List the dimensions of all drip caps to be installed on a building.

- Describe the support for masonry veneer.

- List locations where bricks are to be used in stretcher, header, soldier, and rowlock positions.

## ASSIGNMENT

Refer to the Lake House drawings (in the packet) to complete the assignment.

1. What material is used for the Lake House siding?

2. How are the outside corners of the Lake House siding finished?

3. In some places, the Lake House siding extends below the wood framing onto the foundation. How is the siding fastened to the foundation?

4. What prevents water from running under the siding at the heads of the Lake House windows?

5. Detail 1/7 of the Lake House shows aluminum screen nailed behind the top edge of the siding. What is the purpose of the opening covered by this screen?

6. Describe one use of aluminum flashing under the siding on the Lake House.

# Decks

## Objectives

After completing this unit, you will be able to perform the following tasks:

○ Explain how a deck is to be supported.

○ Describe how a deck is to be anchored to the house.

○ Locate the necessary information to build handrails on decks.

Wood decks are used to extend the living area of a house to the outdoors. A deck may be a single-level platform, or it may be a complex structure with several levels and shapes. However, nearly all wood decks are made of wood planks laid over joists or beams, **Figure 27–1.** The planks are laid with a small space between them, so rainwater does not collect on the deck.

The same construction methods are used for decks and porches as for other parts of the house. The parts of deck construction that require special attention or that were not covered earlier in the text are discussed here.

## Support

The deck must be supported by stable earth. The support must also extend below the frostline in cold climates. The most common method of support is by wood posts that rest on concrete footings, **Figure 27–2.** All wood used in the construction of a deck should be treated with a chemical to prevent decay. For example, Deck detail 3/6 for the Lake House includes a note that all wood is to be CCA treated. CCA is the abbreviation for *chromated copper arsenate*—a common wood preservative.

**Figure 27–1.** Typical deck construction.

There are a number of products made from combinations of wood fibers and synthetic materials (plastics) that may be used for decks and other outdoor structures. These products, called *composites,* are not affected by weather, moisture, or termites. They can generally be worked with ordinary carpentry tools and fasteners. Of course, as with all products, you should read and follow the manufacturer's instructions.

Decks are usually included on the floor plans for the house. The floor plans show the overall dimensions of the decks and the locations of posts, piers, or other support, **Figure 27–3.** If the decks are complex, they may be shown on a separate plan or detail. On the Lake House, the decks are shown on the floor framing plan.

Typically the posts or piers support a beam or girder, which in turn supports joists. The beam may be solid wood or built-up wood. The joists can be butted against or rested on top of the beam in any of the ways discussed for floor framing.

**Figure 27–2.** Foundation for a typical wood deck.

## Anchoring the Deck

The most common ways of anchoring the deck to the building are shown in **Figure 27–4.** If the deck is at the same level as the house floor, the deck joists can be cantilevered from the floor joists. In this case, blocking is required between the joists. If through bolts are used, the holes are drilled when the deck is built. If the deck is secured with anchor bolts in a concrete foundation, the anchor bolts must be positioned when the foundation is poured.

**Figure 27–3.** Deck shown on floor plan.

INTERIOR SPACE        EXTERIOR

STUD WALL —   — 2X PLATE   — DECK PLANKS

— 2X BLOCKING     **CANTILEVERED JOIST**

— CANTILEVERED JOIST

— WOOD SILL

— CMU FOUNDATION

INTERIOR JOIST —   — JOIST HEADER

— WOOD SILL

— DECK PLANKING

THRU BOLT —    **HEADER W/THRU BOLTS**

— DECK JOIST

— HEADER   — CMU FOUNDATION

INTERIOR JOIST —

— JOIST HEADER

— WOOD SILL

— DECK PLANKING

**HEADER W/ANCHOR BOLT**

ANCHOR BOLT —

— DECK JOIST

CONC FOUNDATION —   — HEADER

**Figure 27–4.** Three methods of anchoring the deck to the house.

The anchor bolts or through bolts hold a joist header. The deck joists either rest on top of the header or butt against it. If the joists butt against the header, they are supported by joist hangers or a ledger strip.

## Railings

Most decks have a railing because they are several feet from the ground. Although metal railings are available in ready-to-install form, the architectural style of most wood decks calls for a carpenter-built wood or composite railing. The simplest type of railing is made of uprights and two or three horizontal rails. The uprights are bolted to the deck frame, and the rails are bolted, screwed, or nailed to the uprights. The style of the railing and the hardware involved are usually indicated on a detail drawing, **Figure 27–5.**

1X6

1X4

2-2X4X3'-4"

1X6 @ 1'-3"
DECK TO ₵

2X10 JOISTS
@ 24"O.C.

2-¾"∅X5"
CARRIAGE BOLTS

5/4 X 6
DECKING

7½"

2"

RAILING DETAIL
SCALE: ½" = 1'-0"

**Figure 27–5.** Railing detail.

## ✓ CHECK YOUR PROGRESS

Can you perform these tasks?

- List the dimensions for the size and location of posts or columns to support a deck or porch.
- Describe the framing for a deck.
- Explain how a deck is to be tied to the building.
- Describe the construction of railings for a deck.
- List the sizes of all parts of a deck.

## ASSIGNMENT

Refer to the Lake House drawings (in the packet) to complete the assignment.

1. What supports the south edge of the decks located outside the Lake House living and dining rooms?

2. How far from the outside of the house foundation is the centerline of these supports?

3. How far apart are these supports?

4. How many anchor bolts are required to fasten both of these decks to the Lake House foundation?

5. What is the purpose of the aluminum flashing shown on Deck detail 3/6?

6. What material is used for the railings on the Lake House south decks?

7. How many lineal feet of horizontal rails are there on these decks?

8. What is the total rise from the lower deck to the higher deck? Which deck is higher?

9. What supports the west edge of the deck between the Lake House kitchen and the garage?

# UNIT

# 28

# Finishing Site Work

## Objectives

After completing this unit, you will be able to perform the following tasks:

O Describe retaining walls, planters, and other constructed landscape features shown on a set of drawings.

O Find the dimensions of paved areas.

O Identify new plantings and other finished landscaping shown on a site plan or landscape plan.

As the exterior of the building is being finished or soon after it is finished, the masons, carpenters, and landscapers begin the finished landscape work. Any constructed features (called *site appurtenances*) are completed first. Then trees and shrubs are planted. Finally, lawns are planted.

## Retaining Walls

Retaining walls are used where sudden changes in elevation are required, **Figure 28–1.** The retaining wall retains, or holds back, the earth. Where the height of the retaining wall is several feet, the earth may put considerable stress on the wall. Therefore, it is important to build the wall according to the plans of the designer. A section through the wall usually is included to show

**Figure 28–1.** Retaining wall on site plan.

the thickness of the wall, its foundation, and any reinforcing steel, **Figure 28–2.** For low retaining walls, the site plan may be the only drawing included.

A low retaining wall is sometimes built around the base of a tree when the finished grade is higher than the natural grade. This retaining wall forms a well around the tree, allowing the roots of the tree to "breathe." An example of a tree that will require a well can be seen in **Figure 28–3,** taken from the Lake House site plan. The 24-inch oak is at an elevation of approximately 333 feet, but the finished grade at this point is 336 feet. Therefore, a well 3 feet deep is required.

**Figure 28–2.** Typical retaining wall construction detail.

## Planters

Planters are sometimes included in the construction of retaining walls or attached to the building. In these cases, the information needed to build the planter is included with the information for the building or retaining wall, **Figure 28–4.** The planter is built right along with the house or retaining wall. If a planter that is separate from other construction is included, it is usually shown with dimension on the site plan. A special section may be included with the details and sections to show how the planter is constructed, **Figure 28–5.**

The planter should be lined with a waterproof membrane, such as polyethylene (common plastic sheeting), or coated with asphalt waterproofing. This keeps the acids and salts in the soil from seeping through the planter and staining it. The planter should also include some way for water to escape. This can be through the bottom or through weep holes. **Weep holes** are openings just above ground level. In cold climates, the planter may be lined with compressible plastic foam. This allows the earth in the planter to expand as it freezes, without cracking the planter. If the planter is to have landscape lighting, automatic watering, etc., additional waterproofing may be required where these utilities penetrate the waterproof membrane.

## Paved Areas

Paved areas on housing sites are drives, walks, and patios. Drives and walks are usually described most fully in the specifications for the project. However, the site plan

includes dimensions and necessary grading information for paved areas, as shown in **Figure 28–3.** These dimensions are usually quite straightforward and easy to understand.

*Patios* are similar to drives and walks in that they are flat areas of paving with easy-to-follow dimensions. They may differ from drives and walks by having different paving materials, such as slate, brick, and flagstone, for example. Patios may also be made of a concrete slab with different surface material.

## Plantings

Plantings include three types: grass or lawns, shrubs, and trees. On some projects, such items are planted by the owner. When the builder/contractor does the landscaping, the trees and shrubs are planted first; then, the lawns are planted. Some or all of the trees included in the landscape design may have been left when the site was cleared. Any new trees to be planted are shown with a symbol and an identifying note, as shown in **Figure 28–3.** There is no widely accepted standard for the symbols used to represent trees and shrubs. Most architects and drafters use symbols that represent

**Figure 28–3.** Typical site plan. The 24-inch oak tree near the SW corner of the house requires a well. *Courtesy of Robert C. Kurzon.*

FOUNDATION PLAN

FLOOR PLAN

SECTION THRU BUILDING

**Figure 28–4.** Because the planter is a part of the foundation, it is included on the normal drawings for the house. *Courtesy of Home Planners, Inc.*

TOP SOIL (UNCOMPACTED)
BLUE STONE COPING
4" BLUE STONE
GALV. MTL. TIES-ONE PER 4 SQ. FT.
1" AIR SPACE
4" C.M.U.
ASPHALT COATING
1" PLASTIC FOAM BD.
2" FLAGSTONE (PATIO)
4" CONC. W/8X8 W1.4 X W1.4
4" CLEAN GRAVEL
8" C.M.U.
3'-4" X 10" CONC. FDN.

½" COPPER TUBE @ 2'-0" O.C.

3"
2'-8"  4 COURSES
2'-4"
2'-0"  3 COURSES

**Figure 28–5.** Section through planter and patio.

DECIDUOUS (LEAF SHEDDING) TREES     OR          CONIFEROUS (EVERGREEN) TREES

PALM TREES          LOW SHRUBS          SPREADING GROUND COVER

**Figure 28–6.** Typical symbols for plantings.

deciduous (leaf shedding) trees, coniferous (evergreen) trees, palms, and low shrubs, **Figure 28–6.** The trees and shrubs may also be listed in a schedule of plantings, **Figure 28–7.**

Grass is planted by seeding or sodding. Although a note on the site plan may indicate seeded or sodded areas, more detailed information is usually given on the schedule of plantings or in the specification.

LANDSCAPE PLANTING PLAN FOR:
MR. & MRS. C. BURTON
KENNEDY ROAD
COBLESKILL, N.Y.

LANDSCAPE DESIGN BY DAN PIERRO

# PLANT LIST

| BOTANICAL NAME | COMMON NAME | NUMBER USED |
|---|---|---|
| | ~TREES~ | |
| ACER PALMATUM | JAPANESE RED MAPLE | 1 |
| ACER SACCHARUM | SUGAR MAPLE | 1 |
| BETULA PAPYRIFERA | CLUMP WHITE BIRCH | 1 |
| GLEDITSIA TRIACANTHOS INERMIS | SHADEMASTER HONEY LOCUST | 1 |
| | ~SHRUBS~ | |
| AZALEA MOLLIS | CHINESE AZALEA | 5 |
| COTONEASTER HORIZONTALIS | ROCK SPRAY COTONEASTER | 5 |
| FORSYTHIA INTERMEDIA | FORSYTHIA LYNWOOD GOLD | 31 |
| ILEX CRENATA | JAPANESE HOLLY | 2 |
| ILEX CRENATA CONVEXA BULLATA | CONVEX LEAF HOLLY | 11 |
| JUNIPERUS CHINENSIS ESPALIER | ESPALIER JUNIPER | 1 |
| JUNIPERUS CHINENSIS PARSONI | PARSONI JUNIPER | 18 |
| JUNIPERUS CHINENSIS PFITZERIANA | PFITZER JUNIPER | 13 |

| BOTANICAL NAME | COMMON NAME | NUMBER USED |
|---|---|---|
| | ~TREES~ | |
| JUNIPERUS CHINENSIS TORULOSA | TORULOSA JUNIPER | 1 |
| JUNIPERUS CHINENSIS WILTONI | WILTONI JUNIPER | 7 |
| KALMIA LATIFOLIA | MOUNTAIN LAUREL | 3 |
| PICEA ABIES NIDIFORMIS | BIRD'S NEST SPRUCE | 3 |
| PICEA GLAUCA ALBERTIANA | ALBERTA SPRUCE | 1 |
| PIERIS JAPONICA | ANDROMEDA | 1 |
| PINUS MUGO MUGHUS | MUGO PINE | 1 |
| PINUS THUNBERGI | JAPANESE BLACK PINE | 2 |
| RHODODENDRON CAROLINA | RHODODENDRON CAROLINA | 11 |
| TAXUS CUSPIDATA DENSIFORMIS | DENSIFORMIS JAPANESE YEW | 5 |
| TSUGA CANADENSIS | AMERICAN HEMLOCK | 8 |
| | ~GROUND COVERS~ | |
| VINCA MINOR | MYRTLE | 1200 |

**Figure 28–7.** A landscape plan complete with plant list.

**Figure 28–8.** Site plan for assignment questions.

## ✓ CHECK YOUR PROGRESS

Can you perform these tasks?

● List the dimensions of a retaining wall.

● Describe the footing for a retaining wall.

● Describe provisions for drainage through a retaining wall.

● List the dimensions of a planter.

● Explain how the inside of a planter is to be waterproofed.

● List the dimensions of all paved areas.

● Describe the pitch of all paved areas.

● Name the types of plantings shown on a site or landscape plan.

## ASSIGNMENT

Refer to **Figures 28–8** and **28–9** to complete the assignment.

1. What is the height of the retaining wall above the patio surface at A?

2. How long is the retaining wall?

3. Of what material is the retaining wall constructed?

4. What is the width and what is the length of the patio?

5. What materials are used in the construction of the patio?

6. Describe the weep holes in the planter.

7. How is the planter treated to prevent acids and salts from staining its surface?

8. How many deciduous trees are to be planted?

9. What is the area of the driveway?

10. Assuming that the driveway is 4 inches thick, how many cubic yards of asphalt does it require? (See Math Review 22.)

**Figure 28–9.** Section through planter for assignment questions.

# UNIT

# 29

# Fireplaces

## Objectives

After completing this unit, you will be able to perform the following tasks:

O Describe the foundation, firebox, throat, and chimney of a fireplace using information from a set of construction drawings.

O Explain the finish of the exposed parts of the fireplace, using information from a set of construction drawings.

## Key Terms

O ash dump

O damper

O flue

O hearth

O wythe

## Basic Construction and Theory of Operation of Wood-Burning Fireplaces

A fireplace can be divided into four major parts or zones: foundation, firebox, throat area, and chimney, **Figure 29–1.** Each of these zones has a definite function. To understand the construction details, it is necessary to know how these zones work.

**Figure 29–1.** Four zones of a fireplace.

204   Unit 29

## Foundation

The fireplace foundation serves the same purpose as the foundation of the house—it supports the upper parts and spreads the weight over an area of stable earth. The foundation consists of a footing and walls capable of supporting the necessary weight, **Figure 29–2.** The fireplace foundation sometimes houses an ash pit. The *ash pit* is a reservoir to hold ashes that are dropped through an **ash dump** (a small door) in the floor of the fireplace, **Figure 29–3.** When the foundation includes an ash pit, a *cleanout door* is installed near the bottom of the ash pit, **Figure 29–4.**

## Firebox

The firebox is the area where the fire is built. In all masonry fireplaces, the firebox is constructed of two layers of masonry, as shown in **Figure 29–2.** Each layer is called a **wythe.** The floor of the firebox consists of firebricks laid over a concrete base. The concrete base may extend beyond the face of the firebox to support the hearth. The **hearth,** which may be tile, stone, brick, or slate, forms a noncombustible floor area in front of the fireplace. The outer walls of the firebox are most often common brick. The inner walls are of firebrick to withstand the heat of the fire. The back wall of the firebox slopes in to direct

CAP
FLASHING
FIRESTOP
FLUE LINING
SMOKE CHAMBER
MANTEL
SMOKE SHELF
LINTEL
DAMPER
2 1/2" FIREBRICK
ASH DUMP
BRICK OUTER HEARTH
4" REINFORCED CONCRETE SLAB
ASH PIT
CLEANOUT DOOR
FOOTING
CONCRETE FLOOR
SECTION

**Figure 29–2.** Section view of a masonry fireplace.

**Figure 29–3.** Ash dump. *Courtesy of Majestic Company.*

**Figure 29–4.** Cleanout door. *Courtesy of Majestic Company.*

the smoke and gas into the throat area. The masonry over the fireplace opening is supported by a steel lintel. The lintel is long enough so that 4 inches of it can rest on the masonry at each end.

## Throat

The throat of the fireplace is the area where the firebox narrows into the chimney. Modern wood-burning fireplaces are built with a metal damper in the throat, **Figure 29–5.** The **damper** is a door that can be closed to prevent heat from escaping up the chimney when the fireplace is not in use. The damper is placed on top of the firebox with 1-inch clearance on all four sides. This clearance allows the metal damper to expand as it gets hot.

The flat area behind the damper (above the sloped back of the firebox) is called the *smoke shelf.* The smoke shelf is especially important for the proper operation of the fireplace. The cold air coming down the chimney hits the smoke shelf and turns back up with the rising hot gas and smoke from the firebox. This helps carry the smoke and gas up the chimney, **Figure 29–6.** If the smoke shelf is not built properly and kept clean, the falling cold air can force the smoke and gas back into the firebox.

## Chimney

The chimney carries the smoke and hot gas from the throat to above the house. The top of the chimney must be high enough above the roof, trees, and other nearby obstructions to ensure that the air flows evenly across its top. According to the International Residential Code®, the chimney must extend at least 2 feet through the roof and 2 feet above anything within 10 feet. However, the dimensions on the drawings should always be followed.

To ensure fire safety and a smooth inner surface, masonry chimneys are lined with a clay **flue.** This flue is installed in 2-foot sections as the chimney is built. A 1-inch air space is allowed between the flue lining and the chimney masonry, **Figure 29–7.**

In recent years insulated, metal chimneys have become quite popular. These chimneys are lightweight; they do not require massive foundation for their support. It also takes less time to install them. Because the outer wall of a metal chimney remains cool, it can be enclosed

**Figure 29–5.** Damper. *Courtesy of Majestic Company.*

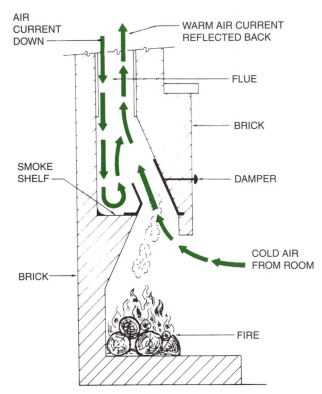

**Figure 29–6.** The smoke shelf turns the incoming air back up the chimney.

**Figure 29–7.** Plan view of a two-flue chimney.

**Figure 29–8.** The framed enclosure for a metal chimney is called a chase.

**Figure 29–9.** Prefabricated metal fireplace. *Courtesy of Majestic Company.*

in wood, **Figure 29–8.** The chimney sections are slipped together and fastened with sheet metal screws; then, the chimney is framed with wood and covered with the specified siding and trim. A chimney enclosure of this type is called a *chase*.

## Prefabricated Metal Fireplaces

Constructing masonry fireplaces is time consuming. Their great weight requires massive foundations. Engineered, metal fireboxes have been developed that can be installed in very little time and require only modest foundations, **Figure 29–9.** The prefabricated units are available from several manufacturers and in a variety of styles. However, they are all similar in that they have double walls and a complete throat with the damper in place. Most also have a firebrick floor.

The double wall improves the heating ability of the metal fireplace. Cool air enters the space between the walls through openings near the bottom. The air absorbs heat from the fire, and because warm air naturally rises, it exits through openings near the top of the unit, **Figure 29–10.** The outside surfaces of the prefabricated unit are cooled by the circulating air; the unit can be enclosed in wood if recommended by the manufacturer.

**Figure 29–10.** In a heat-circulating fireplace, cold air enters the double wall at the bottom, and warm air exits at the top.

For a more traditional appearance, the exposed face of the fireplace can be covered with masonry veneer.

## Gas-Burning Fireplaces

Gas-burning fireplaces are a feature of many new homes and are becoming more common than their wood-burning ancestors. A gas fireplace is an appliance that is installed as a unit. The appliance is installed in a chase, vented to

**Figure 29–11.** Venting through the wall. *Courtesy of Hearth & Home Technologies.*

the outside, and connected to the gas supply. Any style of architectural trim can be applied to the front of the chase after the unit is installed.

Gas fireplaces are efficient, but some heat is lost through the metal surfaces. It is very important that the manufacturer's specifications and instructions be followed to ensure safe venting and to keep combustible materials a safe distance from hot surfaces. Gas-burning fireplaces can be vented straight through the building wall, **Figure 29–11,** or vertically through a chase, **Figure 29–12.** Clearances and framing details are provided in the manufacturer's instructions.

## Fireplace Drawings

Where a gas fireplace is to be installed, most of the information about the fireplace is found in the manufacturer's installation instructions. Gas appliances are fairly lightweight, so there are no special foundation requirements. Wood-burning fireplaces, however, are constructed on

site from masonry materials, so the building plans and elevations must supply all the construction information.

The foundation is normally included on the foundation plan for the building. The dimensions and notes show the location of the fireplace foundation, its size, and the size and type of material to be used. The floor plan of the house shows where the fireplace is located and its overall dimensions. The building elevations show the chimney.

More detailed information about the fireplace is shown on the fireplace details, which usually include a cross section, **Figure 29–13.** The following information is often included on a section view of the fireplace:

O Dimensions of the firebox

O Materials used inside the firebox (firebrick)

O Materials used for the outside of the firebox and chimney

O Ash dump, if any is included

O Lintel over the opening

ATTIC ABOVE

10 IN. (254 MM)

10 IN. (254 MM)

HOLE SHOULD MEASURE
10 IN. X 10 IN.
(254 MM X 254 MM)
INSIDE TO INSIDE

**Figure 29–12.** Venting through a vertical chase. *Courtesy of Hearth & Home Technologies.*

- Dimensions of the hearth
- Mantel, if any is included
- Dimension from the smoke shelf to the flue
- Size of the flue
- Materials for the chimney

An elevation of the fireplace shows the exterior finish of the fireplace, **Figure 29–14.** Only those features that could not be adequately described on the section view are called out on the elevation. However, this view shows the exterior finish—the mantel and the trim—better than the section view does.

FIREPLACE SECTION

SCALE: 3/8 = 1'-0"

12/12 FLUE

1-L 3½ x 3½ x 5/16"

2'-0"

6"

2-L 3½ x 3½ x 5/16"

FIRE BRICK

2'-5"

1'-0"

1'-8"   2'-0"   8"

5 x 8 ASH DUMP

METAL DAMPER

2'-5"

1'-11"

1'-3"

ASH PIT

**Figure 29–13.** A section view shows the construction of the firebox and throat area.

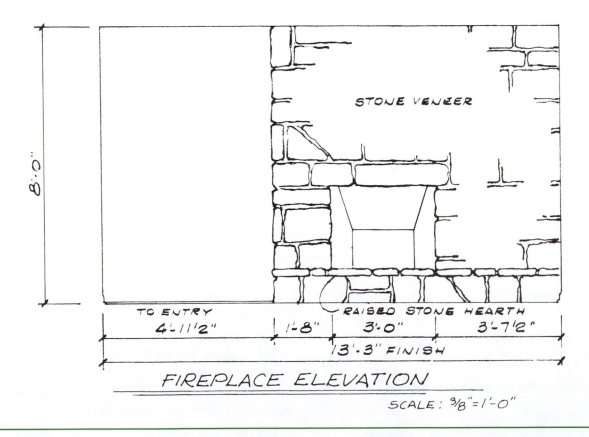

FIREPLACE ELEVATION

SCALE: 3/8" = 1'-0"

8'-0"

STONE VENEER

TO ENTRY
4'-11½"

1'-8"

3'-0"

RAISED STONE HEARTH
3'-7½"

13'-3" FINISH

**Figure 29–14.** An elevation view is a good guide to the finished appearance of the fireplace.

Can you perform these tasks?

● Identify the firebox, smoke shelf, damper, flue, and hearth on a fireplace section.

● List the dimensions of the outside of a fireplace, firebox opening, inside of the firebox, hearth, and height of the chimney for a fireplace.

● Describe the construction, including dimensions, for the enclosure of a prefabricated fireplace.

● Describe the finish and trim for a fireplace.

## ASSIGNMENT

Refer to the Lake House drawings (in the packet) to complete the assignment.

1. What type of fireplace does the Lake House have?

2. How wide is the opening of the firebox?

3. How high is the opening of the firebox?

4. What is the opening next to the fireplace?

5. Determine the overall width and length of the fireplace, including the hearth.

6. Of what material is the hearth constructed?

7. What is used for a lintel over the firebox opening? (Include dimensions.)

8. Briefly describe the foundation of the fireplace.

9. How far above the highest point on the roof is the top of the chimney?

10. What is the total height from the playroom floor to the top of the chimney?

11. What is the overall height of the brickwork involved in the fireplace construction?

12. The top of the fireplace is covered with plastic laminate on ¾-inch plywood. How much clearance is there between that plywood and the chimney?

# Stairs

## Objectives

After completing this unit, you will be able to perform the following tasks:

O Identify the parts of stairs.

O Calculate tread size and riser size.

## Stair Parts

In order to discuss the layout and construction of stairs, you need to know the names of the parts of a set of stairs. The main parts of a stair are defined here and shown in **Figure 30–1.**

O *Stringers* are the main support members. The assembly made up of the stringers and vertical supports is called a **stair carriage.**

O **Treads** are supported by the stringers. The treads are the surfaces one steps on.

O *Risers* are the vertical boards between the treads.

O A *landing* is a platform in the middle of the stairs. Landings are used in stairs that change directions or in very long flights of stairs.

O The *run* of the stairs is the horizontal distance covered by the stairs.

O The *rise* of the stairs is the total vertical dimension of the stairs.

**Figure 30–1.** Basic stair parts.

Figure 30–2. Stair trim and balustrade.

HANDRAIL

NEWEL POST

BALUSTERS

STAIR COVE

NOSING

**Figure 30–3.** Open stringers and open risers.

○ The **nosing** is the portion of the tread that projects beyond the riser.

○ The underside of the nosing may be trimmed with molding called *stair cove.*

○ A *handrail* is usually required on any stairs that are not completely enclosed.

○ **Balusters** are the vertical pieces that support the handrail at each step.

○ The *newel post* is a heavier vertical support used at the bottom of the stair.

○ The balusters, newel post, and handrail together are called a *balustrade.*

The trim parts and balustrade are shown in **Figure 30–2.**

## Types of Stairs

Stairs can be built with open stringers or housed stringers. *Open stringers* are cut in a sawtooth pattern to form a surface for fastening the treads, **Figure 30–3.** *Housed stringers* are routed out, so the treads fit between them, **Figure 30–4.** Stairs are also called open or closed depending on whether the space between the treads is

**Figure 30–4.** Housed stringers and closed risers.

enclosed with risers. Both the kind of stringers and the risers are shown on a section through the stairs. If the stringers are housed, the top of the far stringer shows on the section view. When stair lighting is specified to be installed in the stair riser, the electrical installer must coordinate with the carpenters to avoid a possible conflict with the center or mid-span stringer.

Stairs are also named according to their layout, as seen in a plan view, **Figure 30–5.** Straight stairs are the simplest design. They may or may not include a landing. L-shaped and U-shaped stairs are used where space does not permit a straight run. L-shaped and U-shaped stairs have a landing at the change in directions. The carriages for L-shaped and U-shaped stairs include vertical supports under the landing, **Figure 30–6.** Spiral stairs require the least floor space. They are made up of winder treads, usually supported by a center column.

Temporary stairs used only during construction are usually of cruder construction. They are often made of 2 × 10 stringers with wooden cleats to support the treads. Service stairs in unoccupied spaces, such as a cellar, are often made of 2 × 12 open stringers with 2 × 10 treads. Stairs for exterior decks are also sometimes built with 2 × 12 open stringers, **Figure 30–7.** However, the carpenter should remember that these stairs are important to the overall appearance of the deck.

## Calculating Risers and Treads

Nearly all stair details include a notation showing the number of treads with their width and the number of risers with their height, as shown in **Figure 30–6.** Occasionally only the total run, total rise, number of treads, and number of risers are given on the drawings, as shown in **Figure 30–7.** Notice that there is one more riser than there are treads. The builder must calculate the size of the treads and risers. The steepness of the stairs depends on the relationship between tread width and riser height. If the treads are wide and the risers low, the stairs are gradual. If the treads are narrow and the risers high, the stairs are steep. To climb stairs with wide treads and high risers requires an uncomfortably long stride. To climb stairs with narrow treads and low risers requires an uncomfortably short stride. Building codes specify the maximum height of risers and depth of treads. Section 3.11.5.3 of the International Building Code® specifies that risers cannot be more than 7¾ inches high and treads must be at least 10 inches deep. Although lower risers and deeper treads are allowed by the code, very low risers combined with very deep treads can make for awkward stairs. Stairs are designed by architects to comply with these rules and to be comfortable to climb and descend.

To build the stairs in **Figure 30–7** within these rules, the risers must be 7½ inches high and the treads must be 10 inches wide. The height of the risers can be found by dividing the total rise by the number of risers (3'-1½" or 37½ ÷ 5 = 7½"). (See Math Review 10.) The width of the treads is found by dividing the total run by the number of treads (3'-4" or 40" ÷ 4 = 10").

To calculate the size of the treads and risers in stairs with landings, treat each part as a separate stair. However, the treads and risers should be the same size in each part.

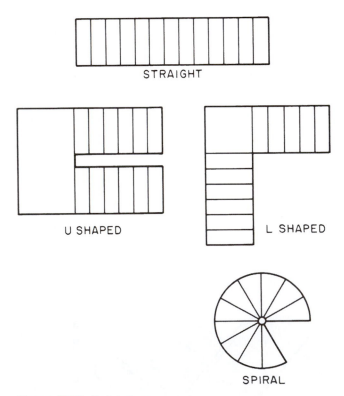

STRAIGHT

U SHAPED

L SHAPED

SPIRAL

**Figure 30–5.** Stair layouts.

6 TREADS @ 10¼" = 5'-1½"

EDGE OF WALL

7 RISERS @ 7½" = 4'-4½"

7 RISERS @ 7½" = 4'-4½"

GALV. JOIST HGR.

1X6

2X8 @ 16" O.C.

½" GYP. BD.

⁵⁄₄ X10 STRINGER

2X8

2X4 PLATE

2-2X10

2X4 @ 24" O.C.

½" GYP. BD.

1⅛" TYP.

⁷⁄₁₆" A.B. @ 2'-0" O.C.

2X4 SOLE PLATE

3'-0"

6 TREADS @ 10¼" = 5'-1½"

STAIR DETAIL

**Figure 30-6.** Stair detail with 2 × 4 studs to support the landing.

STAIR AT DECK

SCALE: $\frac{1}{2}$" = 1'-0"

In the figure: 2 - 2 X 10; 4 TREADS & 5 RISERS; 2 X 10 JOIST; 3'-1$\frac{1}{2}$"; 1" TYP.; 4 X 4 TREATED WOOD POST; 2 X 12 TREATED STRINGER; 3'-4"; FIN. GRADE

**Figure 30–7.** Typical detail for deck stairs.

## ✓ CHECK YOUR PROGRESS

Can you perform these tasks?

● Identify the following stair parts on drawings: stringers, treads, risers, landings, nosing, stair cover molding, newel post, handrail, and baluster.

● Calculate desired tread width and riser heights from the total run and rise.

● List the sizes of all the parts of a stair.

## ASSIGNMENT

Refer to the Lake House drawings (in the packet) to complete the assignment.

Questions 1 through 6 refer to the stairs between the south decks of the Lake House.

1. What is the total run of the stairs between the decks?

2. What is the width of each tread?

3. What is the length of each tread?

4. What is the rise of each step in these stairs?

5. How many stringers are used under these stairs?

6. What size material is used for the stringers?

Questions 7 through 13 refer to the stairs from the kitchen to the bedroom level in the Lake House.

7. How many risers are there, and how high is each riser?

8. How many treads are there, and how wide (front to back including nosing) is each tread?

9. What is the total rise of the stairs?

10. Is this stair built with open or housed stringers?

11. The railing at this stair extends the length of the kitchen hall. If vertical railing supports are spaced at 16" O.C., how many vertical supports are used?

12. How long is each vertical railing support?

13. What material is used for the horizontal rails?

# Insulation and Room Finishing

## Insulation

Thermal insulation is any material that is used to resist the flow of heat. In very warm climates, thermal insulation is used to resist the flow of heat from the outside to the inside. In cold climates, thermal insulation is used to resist the flow of heat from the inside to the outside. Insulating material is rated according to its ability to resist the flow of heat. The measure of this resistance is the R value of the material. The higher the **R value,** the better the material insulates. Typical R values for sidewall or attic insulation range from R-3 to R-38, **Figure 31–1.**

Thermal insulation is generally available in four forms. *Foamed-in-place* materials are synthetic compounds that are sprayed onto a surface and that then produce an insulating foam by a chemical reaction, **Figure 31–2.** *Rigid boards* are plastic foams that have been produced in board form at a factory. Another common type of rigid insulation is made of fiberglass that is manufactured in rigid form instead of as flexible blankets, **Figure 31–3.** Rigid boards are frequently used for foundation insulation, under concrete slabs, and for sheathing. *Blanket insulation* is in the form of flexible rolls or *batts* usually made of fiberglass wool, **Figure 31–4.** Loose *pouring insulation* can be any of a variety of materials that can be poured into place and has good insulating property. A common pouring insulation is loose fiberglass, **Figure 31–5.**

The insulation is shown in the building sections. If the insulation is to be installed between studs, joists, or rafters, it will be sized accordingly. For example, if batts are to be used in a 2 × 4 wall, the insulation will be indicated as 3½ inches thick—the width of a 2 × 4. Where insulation is used in ventilated spaces, there should be room for the necessary air circulation, **Figure 31–6.**

Where the insulation includes a vapor barrier, such as kraft paper, foil, or polyethylene, the vapor barrier is installed on the heated side of the wall. This prevents the moisture in the warm air from passing through the wall and condensing on the cold side of the wall. Such condensation can reduce the R value of the insulation and cause painted surfaces to blister and peel.

## Wall and Ceiling Covering

By far the most widely used wall surface material is **gypsum wallboard.** The most common thicknesses are ⅜ inch, ½ inch, and ⅝ inch. If gypsum board (sometimes called *Sheetrock™*) is to be used over framing that is spaced more

## Objectives

After completing this unit, you will be able to perform the following tasks:

○ Identify the insulation to be used in walls, floors, and ceilings.

○ Identify the wall, ceiling, and floor covering material to be used.

○ List all the kinds of interior molding to be used.

| Type | Typical Thicknesses and R Values | Comments |
|---|---|---|
| Fiberglass blankets & batts | 3/4", R-3<br>2 1/2", R-8<br>3 1/2", R-11<br>3 1/2", R-13<br>3 1/2", R-15<br>6", R-19<br>9", R-30<br>12", R-38 | Flexible blanket-like material. Available with or without vapor barrier on one side |
| Fiberglass blowing wool | R value depends on depth of coverage, but is slightly less than fiberglass blankets. | Other types of blowing or pouring insulation include cellulose and mineral wool. |
| Rigid fiberglass board | 1", R-4.4 | Material is similar to fiberglass blankets, but with rigid binder to create rigid boards. Usually faced with aluminum foil. |
| Rigid urethane foamed board | 1/2", R-3.6<br>3/4", R-5.4<br>1", R-7.2<br>1 1/2", R-10.8<br>2", R-14.6 | Plastic that has been cured in a foamed state to introduce bubbles of air. This creates a rigid foil. |
| Foamed-in-place urethane | R value depends on thickness depth, but is approximately the same as rigid urethane boards. | Other plastic materials may also be foamed in place. |

**Figure 31–1.** Common types of thermal insulation.

**Figure 31–2.** Foamed-in-place insulation. *Courtesy of the Upjohn Company.*

**Figure 31–3.** Rigid insulation. *Courtesy of Owens Corning Fiberglas Corporation.*

than 16 inches O.C. or over masonry, the designer may call for furring. **Furring** consists of narrow strips of wood, usually spaced at 16 or 12 inches O.C., to which the wall covering is fastened, **Figure 31–7.** The furring, if any is to be used, and the thickness of the gypsum wallboard are indicated on the wall sections.

**Figure 31–4.** Blanket insulation. *Courtesy of Owens Corning Fiberglas Corporation.*

**Figure 31–5.** Blowing wool. *Courtesy of Owens Corning Fiberglas Corporation.*

**Figure 31–6.** The insulation must not block the airflow. *Courtesy of Home Planners, Inc.*

*Ceramic tile* also is frequently used on bathroom walls. Ceramic tile may be installed over a base of water-resistant gypsum board, plaster, or plywood.

Gypsum wallboard is the most common ceiling treatment in new home construction. Suspended ceilings are common in commercial construction. Suspended ceilings consist of panels or ceiling tiles supported in a lightweight metal framework. The metal framework is suspended several inches below the ceiling framing on steel wires, **Figure 31–8.**

Building sections or wall sections usually include a typical wall and ceiling. This is representative of what is planned for most of the walls and ceilings in the house. However, there may be some exceptions, such as the

WALLBOARD

FURRING @ 16" O.C.

FRAMING @ 24" O.C.

**Figure 31–7.** Furring is used to provide a nailing surface when framing is spaced too wide or over masonry and concrete.

**Figure 31–8.** Suspended ceiling.

| ROOM | FLOOR | WALLS | CEILING |
|---|---|---|---|
| KITCHEN | QUARRY TILE | GYP. BD. W/WALL PAPER | 12"X12" TILE |
| DINING ROOM | OAK PARQUET | GYP. BD. | GYP. BD. |
| LIVING ROOM | CARPET/PART. BD. | GYP. BD. | GYP. BD. |
| FAMILY ROOM | CARPET/PART. BD. | H.D. BD. PANEL/GYP. BD. | GYP. BD. |
| BEDROOM #1 | CARPET/PART. BD. | GYP. BD. | GYP. BD. |
| BEDROOM #2 | CARPET/PART. BD. | GYP. BD. | GYP. BD. |
| BEDROOM #3 | CARPET/PART. BD. | GYP. BD. | GYP. BD. |
| BATH #1 | CERAMIC TILE | CERAMIC TILE/GYP. BD. | GYP. BD. |
| BATH #2 | CERAMIC TILE | CERAMIC TILE/GYP. BD. | GYP. BD. |
| CLOSETS | CARPET/PART BD. | GYP. BD. | GYP. BD. |
| FOYER | SLATE | GYP. BD. W/WALL PAPER | GYP. BD. |

**Figure 31–9.** Room finish schedule.

bathrooms or kitchen, where water-resistant wallboard is required. Somewhere within the contract documents, you should find a complete list of all room finishes. This may be on one of the drawings, or it may be written into the specifications. **Figure 31–9** shows a room finish schedule that might be included on the drawings. It is common for things like finish color to be left for the owner to choose.

## Finished Floors

A list of possible floor materials would be very long. However, most of the materials fall into one of the following categories: wood, carpet, ceramic, masonry, and resilient materials such as vinyl tile. The finished floor covering is easily found in a schedule of room finishes, but the underlayment for each category is different. **Underlayment** is any material that is used to prepare the **subfloor** to receive the finished floor.

Architectural drafters seem to differ in how they indicate what underlayment is to be used and how the finished floor is to be installed. Some do not include any underlayment on the drawings but rely on the builder's knowledge of good construction practices. This is a dangerous practice. If you find this situation, the architect should be asked for clarification. Sometimes when the drawings do not describe the underlayment, the specifications do. Some drafters indicate the underlayment on the floor plans. Other drafters include the details and section of each area with different types of finished floors, so the underlayment can be shown there.

## Interior Molding

Molding is used to decorate surfaces, protect the edges and corners of surfaces, and conceal joints or seams between surfaces. Molding may be made of wood, plastic, or metal and is available in many shapes and styles. The shapes of commonly used wood molding have been standardized. Each shape is identified by a number, **Figure 31–10.**

Most interior molding is shown on detail drawings. The following are some of the most common uses for interior molding:

- Window casing
- Window **stool**
- Door casing
- Base (bottom of wall)
- **Cove** (top of wall)
- Chair rail (middle of wall)
- Trim around the fireplace mantel
- Trim around built-in cabinets

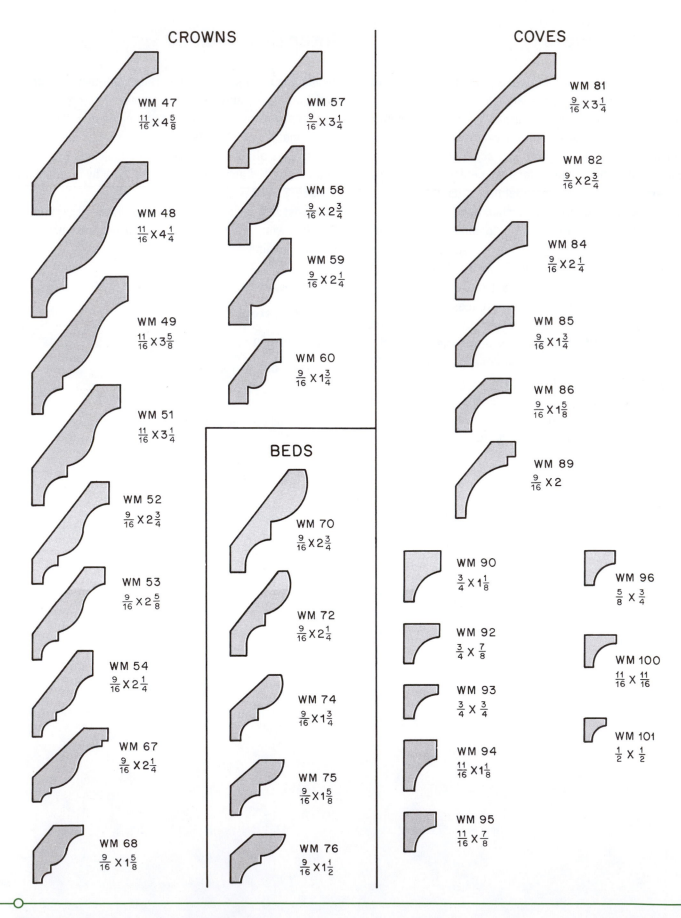

CROWNS

WM 47
$\frac{11}{16} \times 4\frac{5}{8}$

WM 48
$\frac{11}{16} \times 4\frac{1}{4}$

WM 49
$\frac{11}{16} \times 3\frac{5}{8}$

WM 51
$\frac{11}{16} \times 3\frac{1}{4}$

WM 52
$\frac{9}{16} \times 2\frac{3}{4}$

WM 53
$\frac{9}{16} \times 2\frac{5}{8}$

WM 54
$\frac{9}{16} \times 2\frac{1}{4}$

WM 67
$\frac{9}{16} \times 2\frac{1}{4}$

WM 68
$\frac{9}{16} \times 1\frac{5}{8}$

WM 57
$\frac{9}{16} \times 3\frac{1}{4}$

WM 58
$\frac{9}{16} \times 2\frac{3}{4}$

WM 59
$\frac{9}{16} \times 2\frac{1}{4}$

WM 60
$\frac{9}{16} \times 1\frac{3}{4}$

BEDS

WM 70
$\frac{9}{16} \times 2\frac{3}{4}$

WM 72
$\frac{9}{16} \times 2\frac{1}{4}$

WM 74
$\frac{9}{16} \times 1\frac{3}{4}$

WM 75
$\frac{9}{16} \times 1\frac{5}{8}$

WM 76
$\frac{9}{16} \times 1\frac{1}{2}$

COVES

WM 81
$\frac{9}{16} \times 3\frac{1}{4}$

WM 82
$\frac{9}{16} \times 2\frac{3}{4}$

WM 84
$\frac{9}{16} \times 2\frac{1}{4}$

WM 85
$\frac{9}{16} \times 1\frac{3}{4}$

WM 86
$\frac{9}{16} \times 1\frac{5}{8}$

WM 89
$\frac{9}{16} \times 2$

WM 90
$\frac{3}{4} \times 1\frac{1}{8}$

WM 92
$\frac{3}{4} \times \frac{7}{8}$

WM 93
$\frac{3}{4} \times \frac{3}{4}$

WM 94
$\frac{11}{16} \times 1\frac{1}{8}$

WM 95
$\frac{11}{16} \times \frac{7}{8}$

WM 96
$\frac{5}{8} \times \frac{3}{4}$

WM 100
$\frac{11}{16} \times \frac{11}{16}$

WM 101
$\frac{1}{2} \times \frac{1}{2}$

**Figure 31–10.** Common shapes of wood molding.

## QUARTER ROUNDS

WM 103 — $1\frac{1}{16} \times 1\frac{1}{16}$

WM 104 — $\frac{11}{16} \times 1\frac{3}{8}$

WM 105 — $\frac{3}{4} \times \frac{3}{4}$

WM 106 — $\frac{11}{16} \times \frac{11}{16}$

WM 107 — $\frac{5}{8} \times \frac{5}{8}$

WM 108 — $\frac{1}{2} \times \frac{1}{2}$

WM 109 — $\frac{3}{8} \times \frac{3}{8}$

WM 110 — $\frac{1}{4} \times \frac{1}{4}$

## HALF ROUNDS

WM 120 — $\frac{1}{2} \times 1$

WM 122 — $\frac{3}{8} \times \frac{11}{16}$

WM 123 — $\frac{5}{16} \times \frac{5}{8}$

WM 124 — $\frac{1}{4} \times \frac{1}{2}$

## FLAT ASTRAGALS

WM 133 — $\frac{11}{16} \times 1\frac{3}{4}$

WM 134 — $\frac{11}{16} \times 1\frac{3}{8}$

WM 135 — $\frac{7}{16} \times \frac{3}{4}$

## BASE SHOES

WM 126 — $\frac{1}{2} \times \frac{3}{4}$

WM 129 — $\frac{7}{16} \times \frac{11}{16}$

WM 127 — $\frac{7}{16} \times \frac{3}{4}$

WM 131 — $\frac{1}{2} \times \frac{3}{4}$

## SHELF EDGE / SCREEN MOLD

WM 137 — $\frac{3}{8} \times \frac{3}{4}$

WM 141 — $\frac{1}{4} \times \frac{5}{8}$

WM 138 — $\frac{5}{16} \times \frac{5}{8}$

WM 142 — $\frac{1}{4} \times \frac{3}{4}$

WM 140 — $\frac{1}{4} \times \frac{3}{4}$

WM 144 — $\frac{1}{4} \times \frac{3}{4}$

## GLASS BEADS

WM 147 — $\frac{1}{2} \times \frac{9}{16}$

WM 148 — $\frac{3}{8} \times \frac{3}{8}$

## BASE CAPS

WM 163 — $\frac{11}{16} \times 1\frac{3}{8}$

WM 167 — $\frac{11}{16} \times 1\frac{1}{8}$

WM 164 — $\frac{11}{16} \times 1\frac{1}{8}$

WM 172 — $\frac{5}{8} \times \frac{3}{4}$

WM 166 — $\frac{11}{16} \times 1\frac{1}{4}$

## BRICK MOLD

WM 175 — $1\frac{1}{16} \times 2$

WM 180 — $1\frac{1}{4} \times 2$

WM 176 — $1\frac{1}{16} \times 1\frac{3}{4}$

## DRIP CAPS

WM 187 — $1\frac{11}{16} \times 2$

WM 196 — $\frac{11}{16} \times 1\frac{3}{4}$

WM 188 — $1\frac{1}{16} \times 1\frac{5}{8}$

WM 197 — $\frac{11}{16} \times 1\frac{5}{8}$

## CORNER GUARDS

WM 199 — $1 \times 1$

WM 200 — $\frac{3}{4} \times \frac{3}{4}$

WM 201 — $1\frac{5}{16} \times 1\frac{5}{16}$

WM 204 — $1\frac{5}{16} \times 1\frac{5}{16}$

WM 202 — $1\frac{1}{8} \times 1\frac{1}{8}$

WM 205 — $1\frac{1}{8} \times 1\frac{1}{8}$

WM 203 — $\frac{3}{4} \times \frac{3}{4}$

WM 206 — $\frac{3}{4} \times \frac{3}{4}$

## BATTENS

WM 224 — $\frac{9}{16} \times 2\frac{1}{4}$

WM 229 — $\frac{11}{16} \times 1\frac{5}{8}$

## ROUNDS

WM 232 — $1\frac{5}{8}$

WM 233 — $1\frac{5}{16}$

WM 234 — $1\frac{1}{16}$

## SQUARES

WM 236 — $1\frac{5}{8} \times 1\frac{5}{8}$

WM 237 — $1\frac{5}{16} \times 1\frac{5}{16}$

WM 238 — $1\frac{1}{16} \times 1\frac{1}{16}$

WM 239 — $\frac{3}{4} \times \frac{3}{4}$

**Figure 31–10.** (continued)

# SHINGLE PANEL MOLDINGS

WM 207
$\frac{11}{16} \times 2\frac{1}{2}$

WM 209
$\frac{11}{16} \times 2$

WM 210
$\frac{11}{16} \times 1\frac{5}{8}$

WM 212
$\frac{11}{16} \times 2\frac{1}{2}$

WM 213
$\frac{9}{16} \times 2$

WM 217
$\frac{11}{16} \times 1\frac{3}{4}$

WM 218
$\frac{11}{16} \times 1\frac{1}{2}$

# HAND RAIL

WM 230
$1\frac{1}{2} \times 1\frac{11}{16}$

WM 231
$1\frac{1}{2} \times 1\frac{11}{16}$

WM 240
$1\frac{1}{4} \times 2\frac{1}{4}$

# PICTURE MOLDINGS

WM 273
$\frac{11}{16} \times 1\frac{3}{4}$

WM 276
$\frac{11}{16} \times 1\frac{3}{4}$

# SCREEN/S4S STOCK

WM 241    $1\frac{1}{16} \times 2\frac{3}{4}$
WM 243    $1\frac{1}{16} \times 1\frac{3}{4}$
WM 246    $\frac{3}{4} \times 2\frac{3}{4}$
WM 247    $\frac{3}{4} \times 2$
WM 248    $\frac{3}{4} \times 1\frac{3}{4}$
WM 249    $\frac{3}{4} \times 1\frac{5}{8}$
WM 250    $\frac{3}{4} \times 1\frac{1}{2}$
WM 251    $\frac{3}{4} \times 1\frac{3}{8}$
WM 252    $\frac{3}{4} \times 1\frac{1}{4}$
WM 254    $\frac{1}{2} \times \frac{3}{4}$

# LATTICE

WM 265    $\frac{9}{32} \times 1\frac{3}{4}$
WM 266    $\frac{9}{32} \times 1\frac{5}{8}$
WM 267    $\frac{9}{32} \times 1\frac{3}{8}$
WM 268    $\frac{9}{32} \times 1\frac{1}{8}$

# BACK BANDS

WM 280
$\frac{11}{16} \times 1\frac{1}{16}$

WM 281
$\frac{11}{16} \times 1\frac{1}{8}$

# WAINSCOT/PLY CAP MOLDINGS

WM 290
$\frac{11}{16} \times 1\frac{3}{8}$

WM 292
$\frac{9}{16} \times 1\frac{1}{8}$

WM 294
$\frac{11}{16} \times 1\frac{1}{8}$

WM 295
$\frac{1}{2} \times 1\frac{1}{4}$

WM 296
$\frac{3}{4} \times \frac{3}{4}$

# CHAIR RAILS

WM 297
$\frac{11}{16} \times 3$

WM 298
$\frac{11}{16} \times 2\frac{1}{2}$

WM 300
$1\frac{1}{16} \times 3$

WM 303
$\frac{9}{16} \times 2\frac{1}{2}$

WM 304
$\frac{1}{2} \times 2\frac{1}{4}$

WM 390
$\frac{11}{16} \times 2\frac{5}{8}$

# FLAT STOOLS

WM 1021 $\frac{11}{16} \times$ WIDTH SPECIFIED

# T-ASTRAGALS

WM 1300
$1\frac{1}{4} \times 2\frac{1}{4}$

WM 1305
$1\frac{1}{4} \times 2$

WM 1310
$1\frac{1}{4} \times 2\frac{1}{4}$

WM 1315
$1\frac{1}{4} \times 2$

**Figure 31–10.** (continued)

# CASING

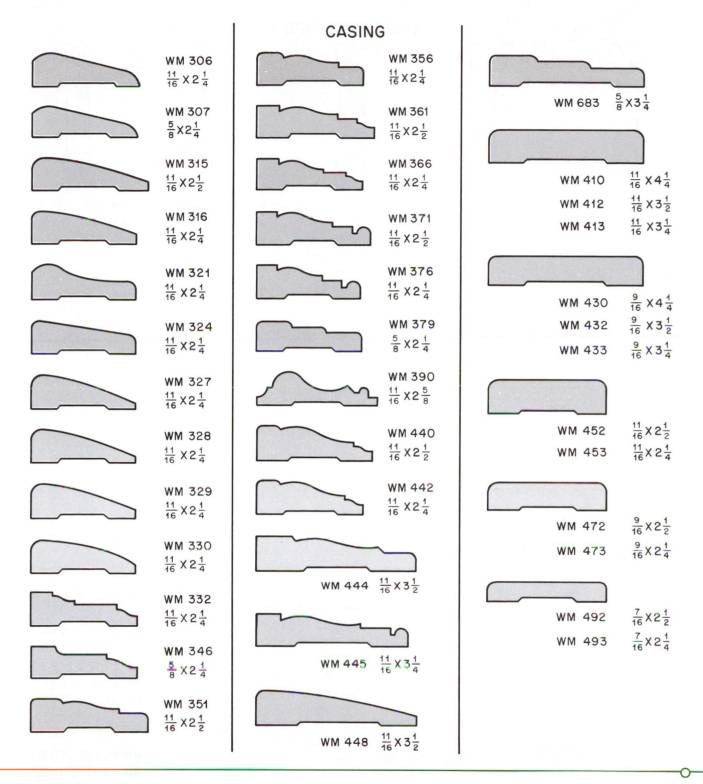

WM 306   $\frac{11}{16}$ X $2\frac{1}{4}$

WM 307   $\frac{5}{8}$ X $2\frac{1}{4}$

WM 315   $\frac{11}{16}$ X $2\frac{1}{2}$

WM 316   $\frac{11}{16}$ X $2\frac{1}{4}$

WM 321   $\frac{11}{16}$ X $2\frac{1}{4}$

WM 324   $\frac{11}{16}$ X $2\frac{1}{4}$

WM 327   $\frac{11}{16}$ X $2\frac{1}{4}$

WM 328   $\frac{11}{16}$ X $2\frac{1}{4}$

WM 329   $\frac{11}{16}$ X $2\frac{1}{4}$

WM 330   $\frac{11}{16}$ X $2\frac{1}{4}$

WM 332   $\frac{11}{16}$ X $2\frac{1}{4}$

WM 346   $\frac{5}{8}$ X $2\frac{1}{4}$

WM 351   $\frac{11}{16}$ X $2\frac{1}{2}$

WM 356   $\frac{11}{16}$ X $2\frac{1}{4}$

WM 361   $\frac{11}{16}$ X $2\frac{1}{2}$

WM 366   $\frac{11}{16}$ X $2\frac{1}{4}$

WM 371   $\frac{11}{16}$ X $2\frac{1}{2}$

WM 376   $\frac{11}{16}$ X $2\frac{1}{4}$

WM 379   $\frac{5}{8}$ X $2\frac{1}{4}$

WM 390   $\frac{11}{16}$ X $2\frac{5}{8}$

WM 440   $\frac{11}{16}$ X $2\frac{1}{2}$

WM 442   $\frac{11}{16}$ X $2\frac{1}{4}$

WM 444   $\frac{11}{16}$ X $3\frac{1}{2}$

WM 445   $\frac{11}{16}$ X $3\frac{1}{4}$

WM 448   $\frac{11}{16}$ X $3\frac{1}{2}$

WM 683   $\frac{5}{8}$ X $3\frac{1}{4}$

WM 410   $\frac{11}{16}$ X $4\frac{1}{4}$
WM 412   $\frac{11}{16}$ X $3\frac{1}{2}$
WM 413   $\frac{11}{16}$ X $3\frac{1}{4}$

WM 430   $\frac{9}{16}$ X $4\frac{1}{4}$
WM 432   $\frac{9}{16}$ X $3\frac{1}{2}$
WM 433   $\frac{9}{16}$ X $3\frac{1}{4}$

WM 452   $\frac{11}{16}$ X $2\frac{1}{2}$
WM 453   $\frac{11}{16}$ X $2\frac{1}{4}$

WM 472   $\frac{9}{16}$ X $2\frac{1}{2}$
WM 473   $\frac{9}{16}$ X $2\frac{1}{4}$

WM 492   $\frac{7}{16}$ X $2\frac{1}{2}$
WM 493   $\frac{7}{16}$ X $2\frac{1}{4}$

**Figure 31–10.** (continued)

## BASE MOLDINGS

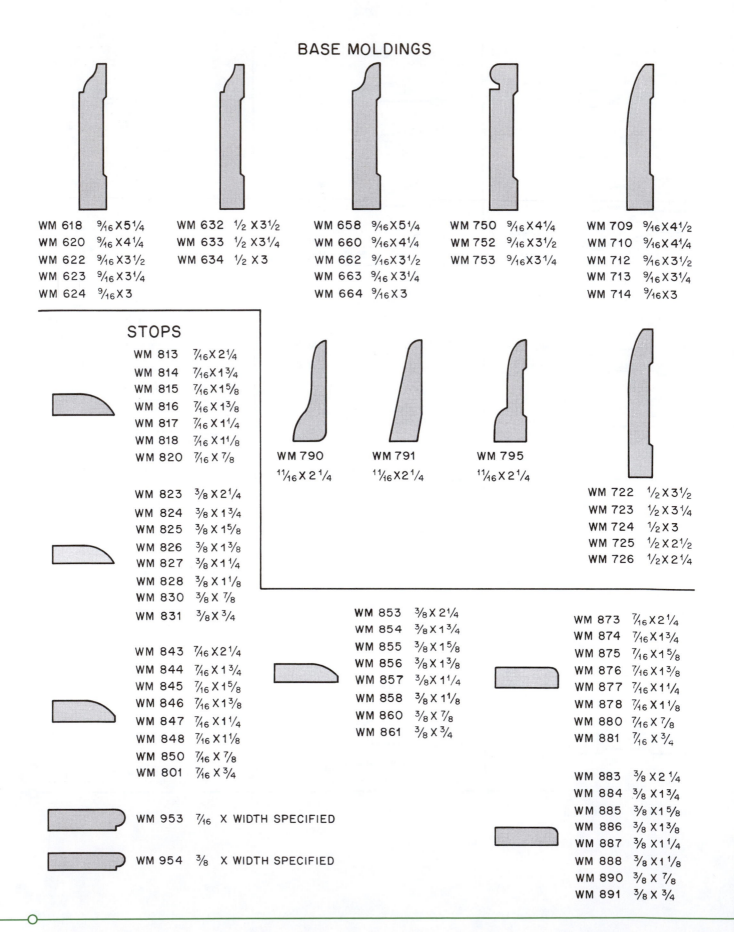

| WM 618 | $9/16 \times 5 1/4$ | WM 632 | $1/2 \times 3 1/2$ | WM 658 | $9/16 \times 5 1/4$ | WM 750 | $9/16 \times 4 1/4$ | WM 709 | $9/16 \times 4 1/2$ |
|---|---|---|---|---|---|---|---|---|---|
| WM 620 | $9/16 \times 4 1/4$ | WM 633 | $1/2 \times 3 1/4$ | WM 660 | $9/16 \times 4 1/4$ | WM 752 | $9/16 \times 3 1/2$ | WM 710 | $9/16 \times 4 1/4$ |
| WM 622 | $9/16 \times 3 1/2$ | WM 634 | $1/2 \times 3$ | WM 662 | $9/16 \times 3 1/2$ | WM 753 | $9/16 \times 3 1/4$ | WM 712 | $9/16 \times 3 1/2$ |
| WM 623 | $9/16 \times 3 1/4$ | | | WM 663 | $9/16 \times 3 1/4$ | | | WM 713 | $9/16 \times 3 1/4$ |
| WM 624 | $9/16 \times 3$ | | | WM 664 | $9/16 \times 3$ | | | WM 714 | $9/16 \times 3$ |

## STOPS

| WM 813 | $7/16 \times 2 1/4$ |
|---|---|
| WM 814 | $7/16 \times 1 3/4$ |
| WM 815 | $7/16 \times 1 5/8$ |
| WM 816 | $7/16 \times 1 3/8$ |
| WM 817 | $7/16 \times 1 1/4$ |
| WM 818 | $7/16 \times 1 1/8$ |
| WM 820 | $7/16 \times 7/8$ |

WM 790  $11/16 \times 2 1/4$

WM 791  $11/16 \times 2 1/4$

WM 795  $11/16 \times 2 1/4$

| WM 823 | $3/8 \times 2 1/4$ |
|---|---|
| WM 824 | $3/8 \times 1 3/4$ |
| WM 825 | $3/8 \times 1 5/8$ |
| WM 826 | $3/8 \times 1 3/8$ |
| WM 827 | $3/8 \times 1 1/4$ |
| WM 828 | $3/8 \times 1 1/8$ |
| WM 830 | $3/8 \times 7/8$ |
| WM 831 | $3/8 \times 3/4$ |

| WM 722 | $1/2 \times 3 1/2$ |
|---|---|
| WM 723 | $1/2 \times 3 1/4$ |
| WM 724 | $1/2 \times 3$ |
| WM 725 | $1/2 \times 2 1/2$ |
| WM 726 | $1/2 \times 2 1/4$ |

| WM 843 | $7/16 \times 2 1/4$ |
|---|---|
| WM 844 | $7/16 \times 1 3/4$ |
| WM 845 | $7/16 \times 1 5/8$ |
| WM 846 | $7/16 \times 1 3/8$ |
| WM 847 | $7/16 \times 1 1/4$ |
| WM 848 | $7/16 \times 1 1/8$ |
| WM 850 | $7/16 \times 7/8$ |
| WM 801 | $7/16 \times 3/4$ |

| WM 853 | $3/8 \times 2 1/4$ |
|---|---|
| WM 854 | $3/8 \times 1 3/4$ |
| WM 855 | $3/8 \times 1 5/8$ |
| WM 856 | $3/8 \times 1 3/8$ |
| WM 857 | $3/8 \times 1 1/4$ |
| WM 858 | $3/8 \times 1 1/8$ |
| WM 860 | $3/8 \times 7/8$ |
| WM 861 | $3/8 \times 3/4$ |

| WM 873 | $7/16 \times 2 1/4$ |
|---|---|
| WM 874 | $7/16 \times 1 3/4$ |
| WM 875 | $7/16 \times 1 5/8$ |
| WM 876 | $7/16 \times 1 3/8$ |
| WM 877 | $7/16 \times 1 1/4$ |
| WM 878 | $7/16 \times 1 1/8$ |
| WM 880 | $7/16 \times 7/8$ |
| WM 881 | $7/16 \times 3/4$ |

WM 953  $7/16$ X WIDTH SPECIFIED

WM 954  $3/8$ X WIDTH SPECIFIED

| WM 883 | $3/8 \times 2 1/4$ |
|---|---|
| WM 884 | $3/8 \times 1 3/4$ |
| WM 885 | $3/8 \times 1 5/8$ |
| WM 886 | $3/8 \times 1 3/8$ |
| WM 887 | $3/8 \times 1 1/4$ |
| WM 888 | $3/8 \times 1 1/8$ |
| WM 890 | $3/8 \times 7/8$ |
| WM 891 | $3/8 \times 3/4$ |

**Figure 31–10.** (continued)

| | | | | | | |
|---|---|---|---|---|---|---|---|
| WM 903 | $7/16 \times 2 1/4$ | WM 913 | $3/8 \times 2 1/4$ | WM 933 | $7/16 \times 2 1/4$ | WM 943 | $3/8 \times 2 1/4$ |
| WM 904 | $7/16 \times 1 3/4$ | WM 914 | $3/8 \times 1 3/4$ | WM 934 | $7/16 \times 1 3/4$ | WM 944 | $3/8 \times 1 3/4$ |
| WM 905 | $7/16 \times 1 5/8$ | WM 915 | $3/8 \times 1 5/8$ | WM 935 | $7/16 \times 1 5/8$ | WM 945 | $3/8 \times 1 5/8$ |
| WM 906 | $7/16 \times 1 3/8$ | WM 916 | $3/8 \times 1 3/8$ | WM 936 | $7/16 \times 1 3/8$ | WM 946 | $3/8 \times 1 3/8$ |
| WM 907 | $7/16 \times 1 1/4$ | WM 917 | $3/8 \times 1 1/4$ | WM 937 | $7/16 \times 1 1/4$ | WM 947 | $3/8 \times 1 1/4$ |
| WM 908 | $7/16 \times 1 1/8$ | WM 918 | $3/8 \times 1 1/8$ | WM 938 | $7/16 \times 1 1/8$ | WM 948 | $3/8 \times 1 1/8$ |
| WM 910 | $7/16 \times 7/8$ | WM 920 | $3/8 \times 7/8$ | WM 940 | $7/16 \times 7/8$ | WM 950 | $3/8 \times 7/8$ |
| WM 911 | $7/16 \times 3/4$ | WM 921 | $3/8 \times 3/4$ | WM 941 | $7/16 \times 3/4$ | WM 951 | $3/8 \times 3/4$ |

## PANEL STRIPS
## MULLION CASINGS

WM 955 $3/8 \times 2 1/4$
WM 956 $3/8 \times 2$
WM 957 $3/8 \times 1 3/4$

WM 960 $3/8 \times 2 1/4$
WM 962 $3/8 \times 2$
WM 963 $3/8 \times 1 3/4$

WM 965 $3/8 \times 2 1/4$
WM 967 $3/8 \times 2$
WM 968 $3/8 \times 1 3/4$

WM 970 $3/8 \times 2 1/4$
WM 972 $3/8 \times 2$
WM 973 $3/8 \times 1 3/4$

WM 975 $3/8 \times 2 1/4$
WM 977 $3/8 \times 2$
WM 978 $3/8 \times 1 3/4$

WM 980 $3/8 \times 2 1/4$
WM 982 $3/8 \times 2$
WM 983 $3/8 \times 1 3/4$

WM 985 $3/8 \times 2 1/4$
WM 987 $3/8 \times 2$
WM 988 $3/8 \times 1 3/4$

## RABBETED STOOLS

SPECIFY WIDTH OF RABBET
AND DEGREE OF BEVEL

WIDTH
BEVEL
10°   14°

WM 1131 $1 1/16 \times 3 5/8$
WM 1133 $1 1/16 \times 3 1/4$
WM 1134 $1 1/16 \times 2 3/4$

WM 1153 $11/16 \times 3 1/4$
WM 1154 $11/16 \times 2 3/4$
WM 1155 $11/16 \times 2 1/2$
WM 1156 $11/16 \times 2 1/4$

WM 1161 $1 1/16 \times 3 5/8$
WM 1163 $1 1/16 \times 3 1/4$
WM 1164 $1 1/16 \times 2 3/4$

WM 1193 $11/16 \times 3 1/4$
WM 1194 $11/16 \times 2 3/4$
WM 1195 $11/16 \times 2 1/2$
WM 1196 $11/16 \times 2 1/4$

**Figure 31–10.** (continued)

Can you perform these tasks?

● Describe the kinds of insulation to be used in walls, around foundations, under slabs, in floors, in attics, and as sheathing.

● List the kind and thickness of material used for the interior surface of walls, ceilings, and floors.

● Identify areas where furring is to be used and the spacing of the furring.

● List all the types of molding to be used for interior trim.

## ASSIGNMENT

Refer to the Lake House drawings (in the packet) to complete the assignment.

1. What size or rating and what kind of insulation are to be used in each of the following locations?
   a. Framed exterior walls
   b. Roof
   c. Under heat sink
   d. Masonry walls of playroom

2. What type of molding is to be used as casing around interior doors?

3. What type of molding is used at the bottom of interior walls and partitions?

4. What kind of trim is used to cover the lower edges of exposed box beams?

5. Describe the wall finish in the playroom including:
   a. What the wall finish material is fastened to
   b. The kind of material used for wall finish

6. What material is used for ceiling finish in the playroom?

7. What material is used for underlayment on typical framed floors?

8. What is the finished floor material at the heat sink?

9. What covers the interior faces of wood box beams?

10. What is the finished wall material in the bedrooms?

# Cabinets

## Showing Cabinets on Drawings

The layout of the cabinets can be determined by reading the floor plan and cabinet elevations together. The floor plan normally includes reference marks indicating each cabinet elevation, **Figure 32–1.** The floor plan shows the location of major appliances and may include some overall layout dimensions.

When the project has separate electrical drawings, the electrical installer must coordinate these designated outlet locations between the electrical and architectural drawings. There may be electrical outlets shown on the architectural drawings and not shown on the electrical drawings. These variances must be resolved early in the rough-in stage of construction.

More complete cabinet information is usually shown on the cabinet elevations, **Figure 32–2.** Cabinet elevations are drawn for each direction from which cabinets can be viewed. These elevations show the types of cabinets, their sizes, and their arrangement. Cabinet types and sizes are recognized by a combination of drawing representations and commonly used letter/number designations. Some drawings rely heavily on standard dimensioning and

## Objectives

After completing this unit, you will be able to perform the following tasks:

O List the sizes and types of cabinets shown on a set of drawings.

O Identify cabinet types and dimensions in manufacturers' literature.

**Figure 32–1.** Kitchen floor plan with key to cabinet elevations.

**Figure 32–2.** Elevations for Figure 32–1. *Courtesy of Home Planners, Inc.*

pictorial representation. This is true of the case shown in **Figure 32–2.** Other drawings include a letter/number designation for each cabinet, as shown in **Figure 32–3.** The letter part of the designation represents the type of cabinet. Architects and drafters vary in how they use letter designations, but they are usually easy to understand after a moment's study. Some typical letter designations are shown in **Figure 32–4.**

The numbers in the cabinet designation represent the dimensions of the cabinet. Base cabinets are a standard

height (usually 3'-0" including the countertop) and a standard front-to-back depth (usually 2'-0") Therefore, base cabinet designations only include two digits to represent width in inches. Wall cabinets are a standard front-to-back depth (usually 1'-0"), but the width and height vary. Therefore, wall cabinet designations include four digits— two for width and two for height. The first and second digits usually indicate width, and the third and fourth digits usually indicate height. The following example explains a typical cabinet designation.

KITCHEN ELEVATIONS

**Figure 32–3.** Letter/number designations for cabinets. *Courtesy of Robert C. Kurzon.*

| LETTER | CABINET TYPE |
|--------|--------------|
| W | WALL CABINET |
| WC | CORNER WALL CABINET |
| B | BASE CABINET |
| D OR DB | DRAWER BASE CABINET |
| BC | BASE CORNER CABINET |
| RC | REVOLVING CORNER CABINET |
| SF | SINK FRONT (ALSO FOR COOKTOP) |
| SB | SINK BASE (ALSO FOR COOKTOP) |
| U | UTILITY OR BROOM CLOSET |
| OV | OVEN CABINET |

**Figure 32–4.** Key to typical cabinet designations.

In addition to cabinets, most cabinet manufacturers provide a variety of accessories:

O Shelves are finished to match the cabinets, so they can be used in open areas.

O Valances are prefinished decorative pieces to use between wall cabinets (over a sink, for example).

O Filler pieces are prefinished boards used to enclose narrow spaces between cabinets.

O A variety of molding may be used for trim.

## Reading Manufacturers' Literature

Cabinet manufacturers publish catalogs and specifications describing their cabinets. The numbering systems used in this literature are similar to those used on construction drawings. There may be slight differences between manufacturers, but these are usually easy to see. **Figure 32–5** is reprinted from one manufacturer's literature. A few tips to reading this manufacturer's literature are listed:

**Figure 32–5.** Cabinet catalog. *Courtesy of H. J. Scheirich Co.*

After three years of intensive research and development the H. J. Scheirich Company is proud to present the new Gardencourt Collection. We wished to make the handsomest cabinets possible — and the toughest . . . cabinets so well styled they would fit into any interior, yet so well built they would pass the most rigorous test of all — everyday use in a busy kitchen.

We discarded all the old woodworking concepts and created a wholly new kind of cabinetry of exceptional structural rigidity and dimensional stability. The basic material for all parts is made of wood fibers bonded solidly together to provide extraordinary strength and resistance to variations in temperature and humidity. Framework, panels, shelves and doors — all are made of the bonded material. Then they are totally sheathed in Vinyl with exterior surfaces in pecan grain finish overlayed with a clear Vinyl for added protection. These are truly "carefree" cabinets. Produced in a modern air-conditioned factory employing highly sophisticated machinery, Gardencourt maintains the well known Scheirich standard of excellence.

All Scheirich cabinets are delivered complete with mounting hardware and detailed instructions for installation. Single door cabinets can hinge left or right. Shipped with doors attached, they can be reversed very simply by moving one hinge screw for each hinge into pre-drilled holes on the opposite inside front of the cabinet frame.

**CORNER WALL cabinets**

C-3024

C-3030, C-3036

C-3042, C-3048

*Corner Wall cabinets are reversible; can be pulled out 6" from corner.*

**CORNER BASE cabinets**

C-427
*Pulls out 9" to 13" from wall*

C-442
*Pulls out 7"*

C-448
*Pulls out 4"*
*Corner Base cabinets arranged for use left or right*

**SINK fronts**

*Sink Front bottom included*

624, 630

636, 642
*Sink fronts above can be cut down 6"*

654
*Can be cut down 12"*

**ALL PURPOSE filler**

Cut on job and assemble as shown. Hardware supplied

AP-42

**END panel**

402

C-3000
**Diagonal Corner Wall cabinet**

C-400
**Revolving Shelf Corner Base cabinet**

**END panels**

601
**SINGLE FACED**
*Use with Sink Front*

602
**DOUBLE FACED**
*Finished Edges*

603
**SINGLE FACED**
*Arranged for L or R*

**VANITIES — GARDENCOURT & GARDENMIST**

**VANITY BOWL bases**

RV-24

RV-30, RV-36

RV-42, RV-48

**VANITY drawer**

VD-36   *Can be cut down 15"*

**TRIM moulding**

**DRAWER VANITY bases**

DV-12
DV-15
DV-18

**ALL PURPOSE filler**

AP-42   *Can be used as Bowl Front*

**VANITY panel**

VP-1
**DOUBLE FACED**
*Finished Edges*

*In keeping with our policy of continuous product refinement and improvement, the H. J. Scheirich Company reserves the right to alter specifications without general notice or obligation to make similar changes in units previously shipped.*

**Figure 32–5.** (continued)

○ Cabinet heights and standard depths are dimensioned on the left side of the page.

○ Referring to these dimensions, you can see that the first and second digits of wall cabinet numbers indicate their height. Therefore, the third and fourth digits indicate width.

○ Standard base cabinets with one drawer above and doors below are designated by three digits only. The first digit (4) indicates a standard base cabinet. The second and third digits indicate width.

○ Base cabinets with other than one drawer above a door are designated by a letter, but the last two digits still indicate width.

## ✓ CHECK YOUR PROGRESS

Can you perform these tasks?

● Orient cabinet elevations to floor plans.

● Describe cabinet types and sizes indicated by typical letter/number designations.

● List all the cabinets shown on the drawings for a house.

## ASSIGNMENT

Refer to the Lake House drawings (in the packet) to complete the assignment.

Using the catalog pages shown in **Figure 32–5** on the following pages, list all the cabinets for the Lake House kitchen.

# TEST

**A. Refer to the Lake House drawings to determine each of the following dimensions.**

1. Northwest corner of the site to the high-water mark at the west boundary

2. Septic tank to the nearest property line

3. Width of the walk on the west side of the house

4. Width of the south end of the drive (in front of the garage)

5. Depth of the earth fill over the septic drainfield

6. Elevation at the butt of the most southerly maple tree to remain

7. Outside of the garage foundation (width × length)

8. Outside of the garage footing (width × length)

9. North to south spacing of the square steel columns

10. Length of the northwest square steel column

11. Length of the haunch under the slab between the north steel columns

12. East-west dimension inside the lower level bathroom at the widest end

13. Closet in bedroom #1 (width × length)

14. Inside of bedroom #1 (width × length)

15. Deck between the kitchen and the garage (width × length)

16. Length of the joists in the loft

17. Height of the foundation wall at the overhead garage door

18. Elevation at the bottom of the concrete piers for the kitchen deck

19. Difference in elevations at the bottom of the south garage footing and the bottom of the nearest house footing

20. Elevation at the bottom of the deepest excavation

21. Width of the cabinet over the refrigerator

22. Width of the cabinet closest to the living room stairs

23. Total thickness of a typical exterior frame wall (allow 1/2 inch for siding)

24. Lineal feet of soldier-course bricks in the fireplace

25. Lineal feet of 2 × 10 lumber in the eave of the corrugated roof

26. Finish floor to the top surface of the freestanding closet/cabinet unit in the kitchen (include surface material)

27. Length of the studs in the wall separating the playroom from the crawl space under the kitchen

28. Length of the steel pins that anchor the foundation walls to the footings

29. Length of the C8 × 11.5 structural steel beam

30. Outside surface of the foundation under the dining room to the centerline of the deck footings

31. Door in the south end of the garage (width × height × thickness)

32. Length of the 2 × 4 studs under the living room bench

33. Width of the treads in the stairs from the kitchen to the bedroom level

34. Rough opening for the door from the hall into bedroom #1 (width × height)

35. Thickness of the concrete at the haunches

B. **Describe the material at each of the following locations in the Lake House. Include such considerations as the kind of material, nominal size of masonry units, and nominal size of lumber.**

1. Tile field pipe

2. Reinforcement in the concrete slabs

3. Rungs in the ladder to loft

4. Hip rafter in the northeast corner of house

5. Roof insulation

6. Roof deck

7. Exterior wall studs

8. Rafter headers at skylight

9. Purlins under the corrugated roof

10. Finished surface on the west wall of playroom

11. Reinforcement in the footings

12. Anchor bolts in the wood sill

13. Wood sill

14. Vapor barrier under the concrete slab

15. Floor underlayment in the bedrooms

16. Floor joists in the kitchen

17. Floor joists in the bedrooms

18. Stair stringers between the decks

19. Railing around the south decks

20. Top course of the foundation at heat sink

21. Expansion joint at the edge of heat sink

22. Rafters above the loft

23. Posts supporting the box beams

24. Girder under the west wall of bedroom #1

25. Base plates on the posts supporting box beams

26. Girder under the kitchen deck joists

27. Finished surface of the living room bench

28. Housed stringer on the north side of stair to bedrooms

29. Stair treads

30. Door casings (interior)

31. Door casings (exterior)

32. Window casings (interior)

33. Glazing (light) in the window on south side of bedroom #1

34. Planks on the kitchen deck

35. Foundation at the west side of the kitchen deck

**C. Answer each of these questions about the Lake House.**

1. How many cubic yards of concrete are required for the concrete slab at elevation 337.00 feet?

2. What is the length of the concrete piers between the kitchen and the garage?

3. How many 4′ × 8′ sheets of plywood are needed to build the box beams?

4. How long are the rafters in the corrugated roof? (Refer to the illustrated rafter table.)

5. How long is the 2 × 12 hip rafter? (Refer to the illustrated rafter table.)

**D. Answer the following questions.**

1. Who should design roof trusses?
   a. architect
   b. carpenter
   c. engineer
   d. any of the above

2. Which drawing will include information about the slope of a roof?
   a. building elevation
   b. truss delivery sheet
   c. truss detail drawing
   d. all of the above

3. What is a piggyback truss?
   a. a truss that is designed to be used as the top half of a large truss
   b. a truss design that uses extra web members
   c. a truss that is to be placed against the side of another truss
   d. a particular brand of roof truss

4. Which sheet would be the easiest one on which to find the required number of trusses of a particular type?
   a. floor plan
   b. truss delivery sheet
   c. truss detail sheet
   d. building elevation

5. What is the most reliable place to find the spacing between trusses?
   a. truss layout
   b. floor plan
   c. truss detail sheet
   d. truss delivery sheet

# MULTIFAMILY CONSTRUCTION

Part III provides an opportunity for you to extend your ability to read construction drawings. Once you have mastered the contents of Parts I and II, you should be able to read and thoroughly understand most residential construction drawings. Part III helps you apply the skills developed earlier to other types of construction and to the work of the mechanical and electrical trades.

The Town House drawings were selected for reference in Part III both because they represent quality construction in a geographic region where many construction practices are different from those in the rest of North America and because they are more complex than earlier drawings, providing an opportunity to sharpen your skills. The Town House involves more floor levels and more complex construction details, so the drawings have more detail information, sometimes making them more difficult to interpret. As you work through these units and the corresponding drawings, take the time to make sure you understand all the information about a particular feature and study all the drawings with information about that part of the building.

The complete set of drawings for the Town House is far too large to include in the packet that accompanies this textbook. Therefore, selected sheets or portions of sheets are used here. To use the available space most efficiently, some of the drawings have been reorganized and combined—with parts of two sheets printed on one. The original sheet and drawing numbers have been retained; therefore, all references printed on the drawings are still applicable.

# Orienting the Drawings

## Objectives

After completing this unit, you will be able to perform the following tasks:

○ Locate a particular building or plan within a large development.

○ Explain the relationships between drawings for construction projects where several plans are to be adjoined in one building.

○ Visualize a building design by reading the drawings.

## Identifying Buildings and Plans

Multifamily dwellings are often built in large developments—with many similar buildings in a single development. Some developments are completed in phases. One phase is completely constructed and begins earning income for the developer before the next stage is started. **Figure 33–1** shows the site plan for Hidden Valley, the development for which the Town House was designed. Hidden Valley is to be developed in four phases. Each phase is outlined with a heavy broken line on the site plan.

The first phase of Hidden Valley includes eleven buildings. Each building is labeled as to building type and the parts that make up the building, **Figure 33–2**. Building Type II is made up of four separate units, each with its own floor plan. Each unit is built like a separate building joined to the next, **Figure 33–3**. Each plan is identified by a letter or letter and numeral. (The term *plan* is used to refer to a particular arrangement of rooms or design. This should not be confused with the use of *plan* to refer to a type of drawing.)

The plans in building Type II are A, B1, B2, and A R. Each plan type is described on separate drawings. Drawings 1 through 5 are for Plan A. Drawings 6 through 10 are for Plan B.

One technique that is used to create similar, yet different, plans is to reverse them. A reversed plan is created by building the plan as though seen in a mirror. The reversed plan has the same features and the same dimensions, but their arrangement is reversed, **Figure 33–4.** In building Type II of the Hidden Valley development, the south end is a reversal of Plan A. This is designated on the site plan and on **Figure 33–2** as A R. The letter *R* indicates a reverse plan.

## Organization of the Drawings

The drawings for a large project must be systematically organized, so information can be found easily. Architects follow similar pattern in organizing their drawings. The general order of drawing sheets is similar to that of specifications: site plans are first; foundation plans and floor plans, next; building elevations, third; then, structural and architectural details; mechanical, next to last; and electrical, last. The cover sheet usually includes an index or table of contents for the drawing set, **Figure 33–5.**

**Figure 33–1.** Site plan. *Courtesy of Berkus-Group Architects.*

The index for the Hidden Valley drawing does not indicate mechanical or electrical drawings. In some parts of the United States, it is common practice not to prepare separate drawings for mechanical or electrical work on town houses. The essential information for these trades is included on the drawings for the other trades.

## Visualizing the Plan

The first step in becoming familiar with any plan should be to mentally walk through the plan. This technique is described in Unit 18 for the Lake House. Refer to the Hidden Valley drawings as you mentally walk through Plan A of the Town House.

Enter through the overhead garage door on the east side of the lowest level. The garage is an open area with stairs leading up to the main floor. Next to the stairs there is an area of dropped ceiling beneath the stairs on the main floor. A note on Drawing 2 refers to Section B.

This note introduces a new consideration for anyone reading construction drawings. Although the architect reviews the drawings carefully, there is always a possibility that errors will appear as one did in this instance. The dropped ceiling beneath the stairs is actually shown on Section E.

**Figure 33–2.** This is a section of the site plan at the size it was drawn. *Courtesy of Berkus-Group Architects.*

A1    B1    B2    A1

**Figure 33–3.** Building Type II includes four housing units. *Courtesy of Berkus-Group Architects.*

The drawings for the Town House are printed exactly as they were prepared for the construction job. A few errors may remain on these drawings. Any errors that do remain give you valuable experience in detecting and dealing with error.

At the top of the garage stairs, there is an entry area with an exterior door. A boxed number 33 refers to the plan notes printed on Drawing 2. This note indicates a stub wall dividing the entry from the dining room. Boxed note references are used throughout the Town House drawings. Reading each related note as it is encountered will help you to better understand the plan.

As you enter the dining room and kitchen, the floor material changes. The kitchen ceiling is dropped to 7'-6"

**Figure 33–4.** A reversed plan is similar to what would be seen by looking at the plan in a mirror.

as dimensioned on Section A-A, Drawing 4. The areas with dropped ceiling above are shown by light cross-hatching on the floor plan. The kitchen has base cabinets on three walls and a peninsula of cabinets separating the kitchen and dining room. Opposite the peninsula of base cabinets, there is a 4'-8" alcove with cabinets, **Figure 33–6.**

Visually walk back through the dining room along a 36"-high guard rail, plan note [13], to a 2-riser stair down to the living room, **Figure 33–7.** At the front of the large 18'-4" × 13'-4" living room there is a fireplace, **Figure 33–8.** To the left of the fireplace is a 7'-0" × 8'-0" sliding glass door that opens onto a small courtyard. A note on the floor plan indicates a *36 MOJ RADIUS ABOVE.* The legend of symbols and abbreviations on sheet O shows that MOJ means *measure on the job.* Looking at the Front Elevation, Drawing 5, you can see that this note refers to a window with a 36-inch radius at the top.

From the entry area, another stair leads up to a hall serving the bedrooms and bath, **Figure 33–9.** Across the hall, a pair of 2'-6" × 6'-8" hollow core doors open

into bedroom #2. By the stairs, a door opens into bathroom #2. This is a small bathroom with a tub, water closet, and lavatory. At the end of the hall, you enter the master bedroom. There is an 8'-0" × 5'-0" aluminum sliding window at the far end of the master bedroom. On your left as you walk into the bedroom, there is a wardrobe closet with a shelf and pole (S&P). On your right there is a master bath. The master bath has a 42" × 60" tub, a vanity (called pullman here), a large wardrobe closet, and a toilet room.

To thoroughly understand the plan, you should read all the plan notes on Drawing A-2 and locate the features to which they refer. Then locate each of the details, indicated by a circle. The numerals above the horizontal line in these circles indicate the detail being referenced. The letter and numeral below the horizontal line indicate on which drawing the detail appears. As you find these plan notes and details, refer to the framing plans, building sections, and elevations to thoroughly understand each.

# table of contents

| | | |
|---|---|---|
| | | COVER SHEET AND INDEX |
| a | | SITE PLAN |
| b | | GENERAL NOTES |
| c | | GENERAL NOTES |
| 1 | PLAN A | FOUNDATION PLAN |
| 2 | | FLOOR PLAN |
| 3 | | FRAMING PLAN |
| 4 | | SECTIONS |
| 5 | | ELEVATIONS |
| 6 | PLAN B | FOUNDATION PLAN |
| 7 | | FLOOR PLAN |
| 8 | | FRAMING PLAN |
| 9 | | SECTIONS |
| 10 | | ELEVATION (1) |
| 11 | | ELEVATION (2) |
| 12 | PLAN C | FOUNDATION PLAN |
| 13 | | FLOOR PLAN |
| 14 | | FRAMING PLAN |
| 15 | | SECTIONS |
| 16 | | ELEVATIONS (1) |
| 17 | | ELEVATIONS (2) |
| 18 | PLAN D | FOUNDATION PLAN |
| 19 | | FLOOR PLAN |
| 20 | | FRAMING PLAN |
| 21 | | SECTIONS |
| 22 | | ELEVATIONS (1) |
| 23 | | ELEVATION (2) |
| 24 | PLAN A-B | INTERIOR ELEVATIONS - PLAN A & B |
| 25 | PLAN C-D. | INTERIOR ELEVATIONS - PLAN C & D |
| 26 | | BUILDING TYPE I |
| 27 | | BUILDING TYPE II |
| 28 | | BUILDING TYPE III |
| 29 | | BUILDING TYPE IV |
| 30 | | BUILDING TYPE V |
| 31 | | BUILDING TYPE VI |
| 32 | | BUILDING TYPE VII |
| 33 | | BUILDING TYPE VII |
| 34 | | BUILDING TYPE VIII |
| 35 | | BUILDING TYPE VIII |
| 36 | | BUILDING TYPE IX |
| 37 | | BUILDING TYPE IX |
| 38 | | BUILDING TYPE X |
| 39 | | BUILDING TYPE X |
| 40 | | BUILDING TYPE XI |
| 41 | | BUILDING TYPE XII |
| 42 | | METER ENCLOSURE |
| 43 | | REC. CENTER    FOUND FLR PLAN |
| 44 | | REC CENTER    SECTION ELEV |
| D1 | | FOUNDATION DETAILS |
| D2 | | FRAMING DETAILS |
| D3 | | FRAMING DETAILS |
| D4 | | FRAMING DETAILS |

*Courtesy of berkus-group architects*

**Figure 33–5.** Table of contents for Hidden Valley construction drawings. *Courtesy of Berkus-Group Architects.*

Figure 33–6. Hidden Valley kitchen as seen from the entry.

Figure 33–8. End wall of living room.

Figure 33–7. Hidden Valley stairs.

**Figure 33–9.** View from Hidden Valley living room.

## ✓ CHECK YOUR PROGRESS

Can you perform these tasks?

● Locate a particular building on the site plan for a large development.

● Locate a particular plan within a multiplan building.

● Locate the drawings for a particular plan type within a large drawing set consisting of several buildings and plan types.

## ASSIGNMENT

Refer to the Town House drawings (including those in the packet and **Figures 33–1** and **33–2**) to complete the assignment.

1. How many buildings are included in phase one?

2. What plans are included in building Type VI?

3. In phase one, Building II, on which side (compass direction) is the courtyard?

4. How thick is the concrete slab for the garage floor in Plan B?

5. What size are the floor joists under the dining room in Plan B?

6. In Plan B, what supports the kitchen floor joists under the back wall of the kitchen?

7. On which drawing would you find elevations of the kitchen cabinets for Plan A? For Plan B?

8. On which drawing would you find details of concrete piers under bearing posts for girders?

9. What is the height of the handrail at the dining room/ living room stairs in Plan B?

10. For each of the major rooms of Plan B listed below, indicate the overall dimensions. Do not include closets, stairs, or minor irregularities. Allow for the thickness of all walls. Walls are dimensioned to the face of the framing.
    a. Living room
    b. Dining room
    c. Kitchen
    d. Deck
    e. Library
    f. Bedroom #2
    g. Master bedroom

11. What are the tread width and the riser height for the stairs between the living room and dining room in Plan B?

12. What important feature of the living room is beside the entry in Plan B?

13. How long are the studs in the partition between the master bedroom and bedroom #2 in Plan B?

14. How high above finished grade is the top of the privacy fence in front of Plan B?

# Town House Construction

## Objectives

After completing this unit, you will be able to perform the following tasks:

O Identify and explain the construction of party walls.

O Find and interpret detailed information on complex residential drawings, such as the Town House.

Multifamily buildings are constructed the same way as single-family buildings in most respects. A wall that is shared by two living units in a multifamily structure is called a *party wall,* **Figure 34–1.** In addition to the usual requirements of a wall, a party wall provides more fire resistance and privacy.

The Town House party walls have stricter fire-resisting requirements in some places than in others. In the Town House drawings, the architect refers to party walls where the fire-resistance factor is lower. Where special fire-code requirements must be met, the architect refers to area-separation walls. The terms *party wall* and *area-separation wall* are often used interchangeably. Other architects may use them with reverse meanings.

**Figure 34–1.** A party wall separates two or more units.

# Fire-Rated Construction

Fire-rated construction serves two purposes in the event of a fire. It slows the spread of fire, and it maintains structural support longer than non-fire-rated construction. Approved fire caulking or fill must be used at all utility-line fire-rated wall penetrations.

The most obvious way in which a wall, floor, ceiling, or roof can slow the spread of fire is by having a fire-resistant surface. Plaster and gypsum wallboard are fire-resistant materials. They do not burn, and they do not easily transmit the heat of fire to the framing members on the other side. Of course, the thicker the wallboard or plaster, the better it resists the flow of heat. For this reason, party walls are often required to have double thicknesses of gypsum wallboard on each side, **Figure 34–2.**

Notice that the wallboard in **Figure 34–2** is indicated as type x. This is a special fire-rated wallboard. Although all gypsum plaster is noncombustible, standard wallboard breaks down and crumbles in the high heat of a fire. Fire-rated wallboard holds up much longer in a fire.

Fire-rated construction is frequently used to separate garages and mechanical rooms from living spaces. Party walls are also fire rated to prevent a fire from spreading between housing units. To completely separate the housing units, a fire-rated party wall should extend all the way from the foundation to the roof.

Most building codes allow an alternative to extending the party wall through the roof. The fire-rated construction may end at the bottom of the roof as long as the roof is of fire-rated construction, **Figure 34–3.** This usually means that the roofing material resists fire for as long as the party wall does. For example, if the party wall is required to be a one-hour code wall (it resists fire for one hour), the roof must be covered only with material that also resists fire for one hour.

Fire-rated walls must also prevent fire from spreading vertically inside the wall. If left open, the spaces between the studs in a frame wall act like chimneys, allowing flame to spread very quickly from one level to another. To prevent vertical flame spread, stud spaces are not permitted to be more than one story high. The spaces between the studs are closed off with **firestops** at each level, **Figure 34–4.** Instead of wood firestops, the wall cavity can be blocked off with fire-resistant fiberglass insulation, as shown in **Figure 34–2.**

**PARTY WALL AT SECOND FLOOR**

SCALE: 1" = 1'-0"

**Figure 34–2.** Party walls often have double layers of gypsum wallboard on one or both sides. *Courtesy of Berkus-Group architects.*

Openings are usually avoided in fire-rated walls. Where it is necessary to include a door, it is made of fire-resistant material. The fire rating of the door must comply with the building code. Fire-rated doors are often allowed to have a slightly lower rating than the walls in which they are installed. Doors in fire walls are equipped with a self-closing mechanism.

## Sound Insulation

To provide privacy between the housing units, party walls should not allow the sound from one unit to be heard in the next unit. The measurement of the capability of a building element to reduce the passage of sound is its *sound transmission classification (STC),* **Figure 34–5.**

Sound is transmitted by vibrations in any material: solid, liquid, or gas. To slow the passage of sound, a party wall must reduce the flow of vibrations. The materials used in the construction of most walls vibrate relatively well. Also, they transmit these vibrations to the air inside the wall. The air carries the sound to the other side, where it is transmitted to the air on the opposite side of the wall. The electrical installer may be required to

FIRE-RATED ROOFING

$\frac{5}{8}"$ FIRE RESISTANT PLYWD.

2 X 10 PLATE

TOP CHORD OF WD. TRUSS

2 X 4 @ 24" O.C. CONT. TO ROOF

ATTIC

2"

$\frac{5}{8}"$ GYP. BD. – TYPE X

2 X 6 NAILER

BOTTOM CHORD OF WD. TRUSS

2 X 10 BLOCKING

$\frac{5}{8}"$ GYP. BD. TYPE X ( TYP.)

ROOM

6" FIBERGLASS BATT INSUL.

SCALE: $\frac{1}{2}"$ = 1'-0"

**Figure 34–3.** Fire-rated party wall in attic space.

FIRE STOPS

**Figure 34–4.** Firestops are installed between studs to prevent vertical drafts inside the wall.

| STC RATING | EFFECTIVENESS |
|---|---|
| 25 | Normal speech can be understood quite easily. |
| 35 | Loud speech can be heard, but not understood. |
| 45 | Must strain to hear loud speech. |
| 48 | Some loud speech can barely be heard. |
| 50 | Loud speech cannot be heard. |

**Figure 34–5.** Sound transmission classes.

utilize fiberglass or some other sound attenuation material around flush device boxes. Raceways may require installations that reduce sound transmission, such as expansion joints, flexible raceway, or raceway offsetting through the party wall.

The sound transmission classification of a wall can be improved greatly by not allowing the studs to contact both surfaces. This may be accomplished in one of two ways. One method is to attach clips made for sound insulation to the studs and then fasten the wallboard to these clips. The clips absorb the vibrations. Using the clips and sound-deadening fiberboard results in an STC rating of 52.

An STC of 45 is achieved without clips by using 2 × 4 studs and 2 × 6 plates. The studs are staggered on opposite sides of the wall so that no studs contact both surfaces, **Figure 34–6**. The STC can be increased to 49 by including fiberglass insulation.

**Figure 34–6.** Typical sound-insulated wall.

## ✓ CHECK YOUR PROGRESS

Can you perform these tasks?

- Identify party walls in a multifamily building.
- Identify fire-rated construction on drawings.
- List the locations of firestops inside framed walls.
- Describe construction to reduce sound transmission.

## ASSIGNMENT

Refer to the Town House drawings (in the packet) to complete the assignment.

1. List the type, thickness, and number of layers of wallboard at each of the following locations in the Town House:
   a. Area separation wall in Plan B dining room
   b. Party wall in Plan B kitchen
   c. Party wall in Plan B garage
   d. Area separation wall in Plan B bedroom #2
   e. Area separation above ceiling in Plan B master bedroom
   f. Party wall in Plan A living room
   g. Garage ceiling under nook in Plan B

2. What is used to stop the vertical spread of fires inside the party walls in the Town House?

3. How is fire prevented from spreading over the top of the area separation walls where they meet the Town House roof?

4. What is the total thickness of the party wall at the library in Plan B?

5. What is done to stop the transmission of sound through the air space in the party wall of the library in Plan B?

6. What is the STC rating of the party wall of the library in Plan B?

7. In Plan B, what is the dimension from the floor to the bottom of the enclosure for the fluorescent light above the master bathroom vanity cabinet and lavatory?

8. On Plan B, on the exterior of the building below the library window there is a sloping shelf. At what angle does that shelf slope?

9. On Plan B, what is the total thickness of the wall separating the master bedroom from the next living unit?

10. What size are the floor joists in the Plan B library?

# UNIT
## 35

# Plumbing, Heating, and Air Conditioning

## Objectives

After completing this unit, you will be able to perform the following tasks:

O  Explain the basic principles of plumbing design.

O  Identify the plumbing symbols used on drawings.

All residences have a plumbing system that consists of a water supply system; a water distribution system; and a drain, waste, and vent (DWV) system. The water supply system provides a source of water to the structure, and the water distribution system provides hot and/or cold water to each of the fixtures requiring water in the structure. After the water is used, the drain, waste, and vent system is used to dispose of the water into a municipal sewage system or an onsite disposal system.

Even though installing gas piping is not defined as plumbing, plumbers sometimes install the piping. The gas piping system supplies fuel gas to the gas-fired fixtures, which might include water heaters, furnaces, ranges, and clothes dryers.

## Plumbing Materials

The materials most often used for plumbing are copper, plastic, cast iron, and black iron. A brief description of each is given in the paragraphs that follow.

*Copper* is frequently used for plumbing because it resists corrosion. However, it is relatively expensive. Copper pipes and fittings may be threaded or smooth for soldered joints.

*Plastic* materials are used in the water supply, water distribution, and DWV systems. These materials are lightweight, noncorrosive, and easily joined. While plastics are allowed in each area of a plumbing system, the plastics used in a water distribution system must have a minimum temperature rating of 180°F (82°C). Plastics are not suitable for some applications where high strength is required.

*Cast iron* is commonly used where DWV piping passes through the foundation and outside the building, where it is buried. Cast iron is strong and has excellent resistance to corrosion. Cast iron is not generally used for water supply or water distribution systems in residential construction.

*Black iron* is used almost exclusively for gas piping. Black iron pipes and fittings are threaded, so they can be screwed together. Brass fittings are frequently used to join black iron pipe.

### Fittings

A wide assortment of fittings is used for joining pipe, making offsets at various angles, controlling the flow of water, and gaining access to the system for service. Most fittings are made of the same materials as the pipe. Plumbers must be familiar with all types of fittings so they can install their work according to the specifications of the designer.

*Couplings,* **Figure 35–1,** are used to join two pipes in a straight line. Couplings are generally used only where a single length of pipe is not long enough.

*Union,* **Figure 35–2,** allows piping to be disconnected without having to cut the pipe. A union consists of three parts, with one part being attached to each pipe and a nut to secure the connection. Then the two parts of the union are screwed together. When it becomes necessary to disconnect the pipe, the two halves of the union are unscrewed.

*Elbows,* **Figure 35–3,** are used to make changes in direction of the piping. Elbows turn either 90°, 45°, or 22.5°.

Some fittings have a hub on each end to accept the outside diameter of the pipe. Others, called street fittings, have a hub on one end, and the other end is the same as the outside diameter of the pipe. Street fittings can be joined directly to other fittings, with no pipe between them.

*Tees and wyes,* **Figure 35–4,** have three connections to allow a second pipe to join the first from the side. Tees have a 90° side connection. A sanitary tee has a curve in the side connection to help direct the flow. Wyes have a 45° side connection.

**Cleanouts, Figure 35–5,** allow access to sewage plumbing for cleaning. A cleanout consists of a threaded opening and a matching plug. When cleaning is necessary, the plug is removed and a drain cleaning cable, also known as a *snake* or *auger,* is run through the line. Cleanouts are installed in each straight run of DWV at the base of drainage stacks, where pipe changes direction more than 45°, and several other areas dictated by plumbing codes.

*Valves* are used to stop, start, or regulate the flow of water. The faucets on a sink or lavatory are a type of

**Figure 35–3.** Street 90° elbow and 45° elbow.

**Figure 35–4.** Sanitary tee and wye.

**Figure 35–1.** A coupling is used to permanently join lengths of pipe.

**Figure 35–2.** A union allows the piping to be disconnected easily.

**Figure 35–5.** A cleanout allows access to the system.

valve. A valve can be used to isolate one part of a system from the rest. Every building must have an isolation valve, and most plumbing fixtures must have an isolation valve located near the fixture. Bathtubs and showers do not typically require separate isolation valves.

## Design of Supply Piping

In most communities, water is distributed through a system of water mains under or near the street. When a new house is constructed, the municipal water department taps (makes an opening in) this main. The supply piping from the municipal tap to the house is installed by plumbers who work for the plumbing contractor.

The main supply pipe entering the house must be larger in diameter than the individual branches installed from the main to each point of use. There are two basic reasons for this. First, water develops friction as it flows through pipes, and the greater size reduces this friction in the long supply line. Second, when more than one fixture is used at a time, the main supply must provide adequate flow for both. Generally, the main supply pipe for a one- or two-family house is ¾-inch or 1-inch pipe.

At the point where the main supply enters the building, a water meter is installed. The water meter measures the amount of water used. The municipal water department relies on this meter to determine the proper water bill for the building. The main water shutoff valve is located near the water meter.

The cold-water distribution piping continues throughout the house, which also provides the water supply to a water heater. The piping system exiting the water heater creates the hot-water distribution system serving the entire house. The size of each pipe providing water to each fixture is dictated by the specific fixture requirements and the plumbing code. **Figure 35–6** shows the water distribution system for a house.

When a valve is suddenly closed at a fixture, the water tends to slam into the closed valve. This causes a sudden pressure buildup in the pipes and may cause the pipes to *hammer* (a sudden shock in the supply piping). For quick-closing valves, water hammer arresters are required by most plumbing codes. The water hammer arrester has a piston that transmits shock waves from the system to a gas-filled chamber, **Figure 35–7**. When a valve is suddenly closed, the gas chamber acts as a shock absorber. Although water cannot be compressed,

gas can be. When the pressure tends to build up suddenly, the gas in the arrester compresses and cushions the resulting shock.

## Drainage Waste and Vent System

The main purpose of a drainage system is to remove wastewater and solids from a building. A drain is installed at each fixture, and all individual drains are connected to eventually create a building drain that exits the building. The building drain connects to a building sewer, which conveys the wastewater and solids to a municipal sewer, septic tank, or other approved point of disposal. The purpose of a vent is to allow air circulation within the system to equalize positive and negative pressures within the piping. Each fixture must be protected by a vent to ensure safe operation of the drainage system. A vent can terminate independently through a roof or be connected with other individual vents to create a branch vent before terminating through a roof. A venting device known as an air admittance valve (AAV) is accepted by some codes. When a plumbing fixture is operated and water drains out of the system, negative pressure causes the valve to open, allowing air to enter as needed to equalize the pressure. When the flow stops, gravity closes the valve, preventing the escape of sewer gases through the valve. An AAV eliminates the need for venting through a roof.

### Traps

The sewer contains foul-smelling, germ-ladened gases that must be prevented from entering the house. If wastewater simply emptied into the sewer from the pipe, this sewer gas would be free to enter the building. To prevent this from happening, a trap is installed at each fixture. A **trap** is a fitting that naturally fills with water to prevent sewer gas from entering the building, **Figure 35–8**. Not all traps are easily seen. Some fixtures, such as **water closets** (toilets), have built-in traps, **Figure 35–9**.

### Vents

As the water rushes through a trap, it is possible for a siphoning action to be started. (The air pressure entering the fixture drain is higher than that on the other side of the trap. This forces the water out of the trap.)

WATER HAMMER
ARRESTERS

WATER HAMMER
ARRESTERS

——————— COLD WATER

━━━━━━━ HOT WATER

**Figure 35–6.** Hot- and cold-water piping.

To prevent DWV traps from siphoning, a vent is installed near the outlet side of the trap. The vent is an opening that allows air pressure to enter the system and break the suction at the trap, **Figure 35–10.** Because the vent allows sewer gas to pass freely, it must be vented to the outside of the building. Unless protected by an AAV, all the fixtures are usually vented into one main vertical pipe, through the roof, **Figure 35–11.**

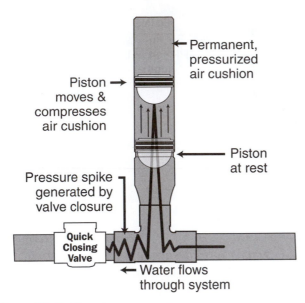

Figure 35–7. Water hammer arrester.

**Figure 35–10.** Venting a trap allows air to enter the system and prevents siphoning.

Figure 35–8. A trap fills with water to prevent sewer gas from entering the building.

## Plumbing Plans

For residential construction, the architect does not usually include a plumbing plan with the set of working drawings. The floor plan shows all the plumbing fixtures by standard symbols. These symbols are easily recognized, because they resemble the actual fixture. The dimensions of the fixtures are provided by the manufacturer on rough-in sheets, **Figure 35–12.**

If the building and the plumbing are fairly simple, plumbers may prepare estimates and bids and complete the work from the symbols on the floor plan only. For more complex houses, the plumbing contractor usually draws a plumbing isometric, **Figure 35–13,** or a special plumbing plan. The drawing set that accompanies this textbook includes a plumbing plan and details for Plan A of the Town House. This sheet includes more details than would normally be found on a plumbing plan for a single-family housing unit. The extra details are included here to help you understand the plumbing plan.

Plumbing plans show each kind of piping by a different symbol. Common plumbing symbols are shown in the Appendix. It will help you understand the plumbing plan if you trace each kind of piping from its source to each fixture. For example, trace the gas piping for the Town House. The gas lines can be recognized by the letter *G* in the piping symbol. The gas supply is shown as a broken

Figure 35–9. A water closet has a built-in trap.

ROOF VENT

VENT FLASHING

VENT LOOP

VANITY

DRUM TRAP

VENT

TO CITY SEWER

3" DWV

CLEANOUT

DRAIN
VENT

WASHING
MACHINE
STANOPIPE

**Figure 35–11.** DWV system.

## ROUGHING-IN DIMENSIONS

| 4035 | 4230 | 4440 |
|------|------|------|

| WATER SEAL | TRAPWAY | WATER SURFACE |
|------------|---------|---------------|
| $3\frac{1}{8}"$ | $2"$ | $12\frac{1}{4}" \times 11\frac{1}{2}"$ |

## SPECIFICATIONS

### FIXTURE SPECIFICATION

☐ **U/R 4035 NEW VENUS** — vitreous china floor-mounted siphon jet close-coupled water-saver combination for 12" rough with regular-rim bowl, requires only 3½ gallons per flush — tank furnished with Fluidmaster 400A ballcock and adjustable tilt flush valve — (2) bolt caps and U/R 8085 lift-off seat included.

### COLOR SPECIFICATION

☐ Acid Resisting White

☐ Acid Resisting Color _____
   (show U/R color desired)

### Universal-Rundle

® 303 NORTH STREET • NEW CASTLE PA 16103

**Figure 35–12.** Typical manufacturer's rough-in sheet. *Courtesy of Universal-Rundle.*

Figure 35–13. Single-line isometric of system shown in Figure 35–12.

Figure 35–14. Sleeve for running gas piping under and through concrete. Local codes specify the design of sleeves used for gas piping. This figure shows only the basic concept.

line until it is inside the garage. Broken lines are used to indicate that the pipe is underground or concealed by construction. Although it is not noted on this plan, the plumbing contractor should know that the building code requires the gas line to be run in a sleeve where it passes through the foundation and the concrete slab, **Figure 35–14.** Just inside the garage wall, the broken line changes to a solid line. At this point, a symbol indicates that the solid line (exposed piping) turns down or away. Here, the gas piping runs above the concrete slab and along the garage wall. A callout on this line indicates that the diameter of the pipe is ¾ inch. At the back of the garage, the gas line has a tee. Both of the outlets of this

tee are ½ inch in diameter. One side of the tee supplies the forced-air unit (F.A.U.). The other side of the tee continues around behind the F.A.U. to another tee and then to the water heater. The side outlet of the second tee supplies a log lighter in the fireplace. This branch is shown on the first floor plumbing plan. Notice that the log-lighter branch is reduced further to ¼-inch diameter.

You should trace each type of piping in a similar manner to be sure you understand it. Using colored pencils to trace the different types of piping may eliminate some of the confusion on crowded drawings.

Refer to the details on the drawing for clarification of the complex areas. As you trace each line, look for the following:

O Kind of plumbing (hot water, cold water, waste)

O Diameter

O Fittings

O Exposed or concealed

O Where line passes through building surfaces

# Heating, Ventilating, and Air Conditioning

Plans for residential construction do not usually include sheets for HVAC (heating, ventilating, and air conditioning), but the floor plan often includes a small amount of basic information—for example, where major equipment such as the furnace, air-handling unit, or air-conditioning unit is to be located and where diffusers (conditioned air outlets) and air returns are to be placed. The HVAC contractor installs the equipment according to building codes and industry practices without the aid of formal drawings.

Many HVAC systems are of the forced-air type. When heat is called for, these systems provide some means of forcing air over a surface heated by gas, oil, or electricity. That surface might be a combustion chamber, a liquid-filled coil, or an electric heating element. When cooling is

called for, the air is forced over a cold coil from a refrigeration unit or a heat pump. The conditioned air is then forced through sheet metal or plastic ducts. The ducts carry the air to openings called diffusers throughout the house. The air is finally picked up by the return grille and returned to the forced-air unit, **Figure 35–15**.

Other types of heating systems include electric resistance heat at the point of use; radiant floor heat, hot water circulated through tubing in or under the floor; and hot water circulated through baseboard heating units. With any of these systems, cooling can only be provided by a separate air-conditioning system.

**Figure 35–15.** The air cycle in a forced-air system.

## ✓ CHECK YOUR PROGRESS

Can you perform these tasks?

- List the plumbing fittings shown on a plumbing plan.
- List the material and size of all piping and fittings for waste plumbing.
- List the material and size of all piping and fittings for supply plumbing.
- Use the manufacturer's literature to list dimensions for the location of plumbing to fixtures.

## ASSIGNMENT

Refer to the Town House Plan A drawings (in the packet) to complete the assignment.

1. What size pipe supplies the washing machine?

2. What size is the cold-water supply to the water heater?

3. What size is the cold-water branch to the lavatory in bathroom #2?

4. At what point does the ¾-inch cold-water branch to the kitchen reduce to ½ inch for the hose bibb?

5. Does the cold-water supply turn up or down as it leaves the bathroom area to supply the kitchen area?

6. List each of the fittings that water will pass through after it drains out of the master bathroom lavatory.

7. List each of the fittings that water will pass through to flow from the main shutoff at the building line to the shutoff on the supply side of the water heater.

8. What size is the waste piping from the water closet in the master bath?

9. What size is the waste piping from the kitchen sink?

10. What size is the waste piping from the washing machine?

11. Where is the air-conditioning compressor located?

12. Where is the tubing that connects the air-conditioning compressor to the forced-air unit?

13. What size tubing connects the compressor to the forced-air unit?

14. How many diffuser outlets supply conditioned air to the rooms?

15. Where are the sizes of the diffuser outlets given?

# UNIT 36

# Electrical

## Objectives

After completing this unit, you will be able to perform the following tasks:

○ Identify the electrical symbols shown on a plan.

○ Explain how the lighting circuits are to be controlled.

## Current, Voltage, Resistance, and Watts

To understand the wiring in a building, you should know how electricity flows. Electricity is energy. To do any work (turn a motor, light a lamp, or produce heat) the electrical energy must have movement. This movement is called *current*. The amount of current is measured in **amperes,** sometimes called *amps.* A single household-type lightbulb requires a current of slightly less than 1 ampere. An electric water heater might require 50 amperes.

The amount of force or pressure causing the current to flow affects the amount of current. The force behind an electric current is called *voltage.* If 115 volts causes a current flow of 5 amperes, 230 volts will cause a current flow of 10 amperes.

The ease with which the current is able to flow through the device also affects the amount of current. The ease or difficulty with which the current flows through the device is called the *resistance* of that device. As the resistance goes up, the current flow goes down. As the resistance goes down, the current flow goes up.

The amount of work the electricity can do in any device depends on both the amount of current (amps) and the force of the current (**volts**). Electrical work is measured in **watts.** The number of watts of power in a device can be found by multiplying the number of amperes by the number of volts. Stated another way, the current flowing in a device can be found by dividing the number of watts by the voltage. For example, how much current flows through a 1,500-watt heater at 115 volts? 1,500 divided by 115 equals about 13 amperes. **Figure 36–1** shows the current, wattage, and voltage of some typical electrical equipment.

| DEVICE | AMPERES | VOLTS | WATTS |
|---|---|---|---|
| Ceiling light fixture | 1.3 | 115 | 150 |
| Vacuum Cleaner | 6.1 | 115 | 700 |
| Radio | 0.4 | 115 | 4 |
| Clock | 0.4 | 115 | 4 |
| Dishwasher | 8.7 | 115 | 1,000 |
| Toaster | 13 | 115 | 1,500 |
| Cook Top | 32 | 230 | 7,450 |
| Oven | 29 | 230 | 6,600 |
| Clothes Dryer | 25 | 230 | 5,750 |
| Washing Machine | 10 | 115 | 1,150 |
| Garbage Disposal | 7.4 | 115 | 850 |

**Figure 36–1.** Current, voltage, and power ratings of some typical electrical devices.

# Circuits

In order for current to flow, it must have a continuous path from the power source, through the electrical device, and back to its source. This complete path is called a *circuit,* **Figure 36–2.**

Many circuits include one or more switches. A switch allows the continuous path to be broken, **Figure 36–3.** By using two 3-way switches, the circuit can be controlled from two places, **Figure 36–4.** When the circuit is broken by a switch or a broken wire, or for any other reason, it is said to be *open.*

Any material that carries electric current is called a **conductor.** In **Figure 36–2** each of the wires is a conductor. When two or more wire conductors are bundled together, they make a cable, **Figure 36–5.**

In larger buildings the wiring is frequently installed by pulling individual wires through steel or plastic pipes,

**Figure 36–2.** A complete circuit includes a path from the supply to the device and back again.

**Figure 36–3.** A switch is used to break (or open) the circuit.

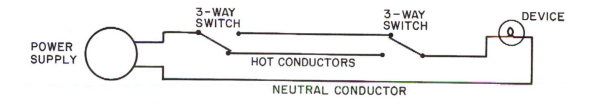

**Figure 36–4.** Three-way switches allow a device to be controlled from two locations. Notice that if either switch is activated, the device will be energized.

**Figure 36–5.** This cable has two circuit conductors and one ground conductor. *Courtesy of Anaconda Wire and Cable Division.*

called *conduit.* In houses it is more common to use cables containing the needed wires plus one ground conductor. The ground conductor does not normally carry current. The *ground,* as it is usually abbreviated, connects all the electrical devices in the house to the ground. If, because of some malfunction, the voltage reaches a part of the device that someone might touch, the ground protects him or her from a serious shock. The current that might otherwise flow through the person follows the ground conductor to the earth. The earth actually carries this current back to the generating station.

Additional protection against serious shock can be provided by using a *ground-fault circuit interrupter* (GFCI or GFI). A GFCI is a device that measures the flow of current in the hot (supply) conductor and the neutral (return) conductor. If a faulty device allows some of the current to flow through a person rather than the neutral conductor, the GFCI stops all current flow immediately.

GFCIs are so effective that the *National Electric Code*® requires their use on circuits for outlets installed outdoors, in kitchens, in bathrooms, in garages, and near any other water hazards.

The electrical service entrance is discussed in Unit 11. The service feeder cable ends at a distribution panel. From the distribution panel, the electrical system is split up into several **branch circuits, Figure 36–6.** Each branch circuit includes a circuit breaker or fuse. The circuit breaker or fuse opens the circuit if the current flow exceeds the rated capacity of the circuit. Branch circuits for special equipment such as water heaters and air conditioners serve that piece of equipment only. Branch circuits for small appliances and miscellaneous use may serve several outlets. Branch circuits for lighting are restricted to lighting only, but a single circuit may serve several lights. Lighting circuits also include switches to turn the lights on and off.

**Figure 36–6.** The electrical service is split up into branch circuits at the distribution panel.

The National Fire Protection Association publishes the *National Electrical Code®*, which specifies the design of safe electrical systems. Electrical engineers and electricians must know this code, which is accepted as the standard for all installations. The following are among the items it covers:

O Kinds and sizes of conductors

O Locations of outlets and devices

O Overcurrent protection (fuses and circuit breakers)

O Number of conductors allowed in a box

O Safe construction of devices

O Grounding

O Switches

The specifications for the structure indicate such things as the type and quality of the equipment to be used, the kind of wiring, and any other information that is not given on the drawings. However, electricians must know the *National Electrical Code®* and any state or local codes that apply because specifications sometimes refer to these codes.

## Electrical Symbols on Plans

The drawings for residential construction usually include electrical information on the floor plans. Only the symbols for outlets, light fixtures, switches, and switch wiring are included. The exact location of the device may not be dimensioned. The position of the device is determined by the electrician after observing the surrounding construction. It should also be noted that all wiring is left to the judgment of the electrician and the regulations of the electrical codes. Switch wiring for light fixtures is included only to show which switches control each light fixture. Switch wiring is shown by a broken line connecting the device and its switch, **Figure 36–7.**

In rooms without a permanent light fixture, one or more convenience outlets may be split wired and controlled by a switch. In split wiring, one-half of the outlet is always hot; the other half can be opened by a switch, **Figure 36–8.** The most common electrical symbols are shown in the Appendix at the back of this book.

**Figure 36–7.** Switch legs on a plan.

ALWAYS HOT

SWITCHED

**Figure 36–8.** Split-wired outlet.

Can you perform these tasks?

● List all the convenience outlets, special outlets, light fixtures, and switches shown on plans.

● Describe the location of the main electrical distribution panel.

## ASSIGNMENT

Refer to the drawings for the Town House, Floor Plan A, including the garage.

1. How many light fixtures are shown on the floor plan? (Include the garage.)

2. How many switches are shown on the floor plan? (Include the garage.)

3. How many duplex outlets are shown?

4. Briefly describe the location of each split-wired outlet and the switch or switches that control each.

5. List five pieces of equipment shown on the floor plan that probably require separate branch circuits.

6. How many outlets are to have ground-fault circuit interrupters included in their circuits?

7. What is the location of the switch or switches that control the light over the stairs to the bedroom level?

8. What is the location of the switch or switches that control the light fixture over the stairs to the garage?

9. Where is the smoke detector located?

10. Where is each of the two telephone outlets?

# TEST

**PART III**

**A. For each of the symbols in Column I, indicate the object in Column II it represents.**

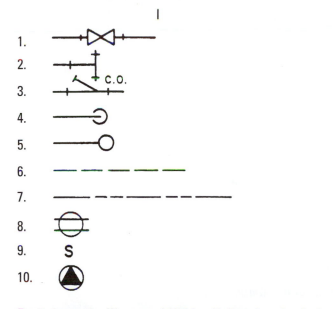

<table>
<tr><td>I</td><td></td><td>II</td></tr>
<tr><td>1.</td><td></td><td>a. Electrical outlet for special equipment</td></tr>
<tr><td>2.</td><td></td><td>b. Pipe turning down or away</td></tr>
<tr><td></td><td>C.O.</td><td>c. Switch wiring</td></tr>
<tr><td>3.</td><td></td><td>d. Convenience outlet</td></tr>
<tr><td>4.</td><td></td><td></td></tr>
<tr><td>5.</td><td></td><td>e. Valve</td></tr>
<tr><td>6.</td><td></td><td>f. Pipe turning up or toward</td></tr>
<tr><td>7.</td><td></td><td>g. Hot-water piping</td></tr>
<tr><td>8.</td><td></td><td>h. Wall switch</td></tr>
<tr><td>9.</td><td>S</td><td>i. Wye fitting with a cleanout</td></tr>
<tr><td>10.</td><td></td><td>j. Tee fitting</td></tr>
</table>

**B. Refer to the illustrated Hidden Valley drawing index and partial site plan to answer the following questions.**

1. In which building is unit 56 located?

2. On which drawing sheet is the foundation plan for unit 56?

3. On which drawing sheet are the foundation details for Plan B units?

4. List all the drawing sheets that are not needed to construct Building IV.

5. Of the buildings shown, which would be built first?

**C. Refer to the Town House drawings to answer these questions.**

1. In Plan B, which rooms have their floors at the same elevation as the kitchen?

2. In Plan B, what separates the living room and dining room?

3. In Plan B, as you climb the stairs from the first floor, what room do you enter?

4. In Plan B, as you climb the stairs from the garage, what room do you enter?

5. In Plan B, what is the distance from the bottom of the garage stairs to the finished floor in the library?

6. In Plan B, what is the width in the living room area, measuring to the inside face of the framing?

7. What size are the anchor bolts in the foundation of Plan B?

8. In the front part of the foundation plan for Plan B, there is a 24"-square by 12"-deep pad for a pier that supports a wood beam. What does this beam support?

9. In Plan B, what supports the floor joists under bedroom #2 at the end nearest the front of the building?

10. In Plan B, what prevents the party wall in the nook area from wracking?

11. In Plan B, where the bedroom floor joists rest on the dining room–kitchen partition, how is the blocking between the joists fastened to the top plate of the wall?

12. What is the wall surface material at the area separation wall in the dining room of Plan B?

13. What is the total thickness, including surface material, of the party wall in the garage stairway in Plan B?

14. What size lumber (thickness $\times$ width $\times$ length) would be ordered for the joists in the Plan A kitchen floor?

15. In Plan A, what is the finished size (thickness $\times$ width $\times$ length) of the floor joists under bedroom #2?

16. What is the length of the cantilevered part of the floor joists under bedroom #2 of Plan A?

17. What material is used for the wood sill on the foundation of Plan A at the exterior corner of the kitchen?

18. In Plan A, what kind and size of fasteners are used to anchor the railing outside the dining room entrance?

19. In Plan A, how far is the air-conditioner compressor pad from the forced-air unit?

20. In Plan B, how many three-way switches are required?

21. In Plan A, how many ceiling light fixtures are there, including luminous soffits?

22. In Plan A, how many plumbing vents penetrate the roof?

23. What diameter of pipe supplies the hose bibb at the front of the kitchen?

24. In Plan A, how many cleanouts are included in the waste piping?

25. In Plan A, where is the hot-water shutoff closest to the shower in bathroom #2?

26. In Plan A, where is the forced-air unit located?

27. How many return air grilles are there on Plan A?

28. What kind of fuel does the F.A.U. use?

29. What is the major difference between units 38 and 42 in Building IV? (Refer to the partial site plan on the facing page.)

30. What is the thickness and depth of the header over the Plan A garage door?

31. How much is the concrete slab thickened for a haunch at the entrance to the garage in Plan A? Give the thickness or width and depth of the haunch without the floor.

# table of contents

| | | |
|---|---|---|
| a | | COVER SHEET AND INDEX |
| b | | SITE PLAN |
| c | | GENERAL NOTES |
| c | | GENERAL NOTES |
| 1 | PLAN A | FOUNDATION PLAN |
| 2 | | FLOOR PLAN |
| 3 | | FRAMING PLAN |
| 4 | | SECTIONS |
| 5 | | ELEVATIONS |
| 6 | PLAN B | FOUNDATION PLAN |
| 7 | | FLOOR PLAN |
| 8 | | FRAMING PLAN |
| 9 | | SECTIONS |
| 10 | | ELEVATION (1) |
| 11 | | ELEVATION (2) |
| 12 | PLAN C | FOUNDATION PLAN |
| 13 | | FLOOR PLAN |
| 14 | | FRAMING PLAN |
| 15 | | SECTIONS |
| 16 | | ELEVATIONS (1) |
| 17 | | ELEVATIONS (2) |
| 18 | PLAN D | FOUNDATION PLAN |
| 19 | | FLOOR PLAN |
| 20 | | FRAMING PLAN |
| 21 | | SECTIONS |
| 22 | | ELEVATIONS (1) |
| 23 | | ELEVATION (2) |
| 24 | PLAN A-B | INTERIOR ELEVATIONS - PLAN A & B |
| 25 | PLAN C-D | INTERIOR ELEVATIONS - PLAN C & D |
| 26 | | BUILDING TYPE I |
| 27 | | BUILDING TYPE II |
| 28 | | BUILDING TYPE III |
| 29 | | BUILDING TYPE IV |
| 30 | | BUILDING TYPE V |
| 31 | | BUILDING TYPE VI |
| 32 | | BUILDING TYPE VII |
| 33 | | BUILDING TYPE VII |
| 34 | | BUILDING TYPE VIII |
| 35 | | BUILDING TYPE VIII |
| 36 | | BUILDING TYPE IX |
| 37 | | BUILDING TYPE X |
| 38 | | BUILDING TYPE X |
| 39 | | BUILDING TYPE X |
| 40 | | BUILDING TYPE X |
| 41 | | BUILDING TYPE XI |
| 42 | | METER ENCLOSURE |
| 43 | | REC. CENTER      FOUND. FLR. PLAN |
| 44 | | REC. CENTER      SECTION ELEV. |
| D1 | | FOUNDATION DETAILS |
| D2 | | FRAMING DETAILS |
| D3 | | FRAMING DETAILS |
| D4 | | FRAMING DETAILS |
| | | |

*Courtesy of berkus-group architects*

# PART

# IV

# HEAVY COMMERCIAL CONSTRUCTION

Part IV presents a thorough examination of the information found on prints for heavy commercial construction. The materials and methods used for large buildings are different from those used in light-frame construction. Also, the drawing set for a commercial project usually includes many more sheets than are found in the drawing sets for smaller buildings. To understand these drawings and make practical use of them, you will need to understand the organization of the drawings and how the heavier materials are described on the drawings.

Part IV also covers air conditioning and heating, plumbing, and electrical drawings in much more detail than earlier sections of the textbook. For those who work in the mechanical or electrical trades, the importance of understanding the drawings for these systems is obvious. However, all the trades work in the same spaces, and the work of one trade affects the other trades. Estimators, superintendents, inspectors, and many other construction professions also require an understanding of all the construction trades.

The School Addition is an excellent building for inclusion in this book. It uses structural steel framing, reinforced concrete foundations, and a varied assortment of other materials typically associated with heavy construction; yet, unlike most large commercial buildings, the School Addition is small enough, so that most of the drawings can be packaged with this book. Some drawings have been eliminated, either because they pertain to other work done on the existing school building or because they do not provide new opportunities for the study of commercial construction drawings. All the School Addition drawings are numbered as they were in the complete set, and the textbook makes frequent reference to the drawing numbers. The Material Keynotes, which are explained in this part of the textbook, are printed on the tabbed pages at the end.

# UNIT 37

# Heavy Commercial Construction

## Objectives

After completing this unit, you will be able to perform the following tasks:

○ Describe the major differences between light-frame construction and heavy commercial construction.

○ Describe common structural steel shapes and their designations.

○ Explain the differences between foundations for heavy structures and those for light-frame buildings.

## Architectural Style

Large buildings, such as the School Addition described in the drawings with this book, are usually designed in a different style from one- and two-family homes or even larger light-frame buildings, such as the Town House.

Most larger commercial buildings are too large to employ the same construction methods as those used with homes and smaller buildings and have different architectural styles than they do, **Figure 37–1.** For example, commercial buildings usually have a larger roof area, so it is not practical to slope the roof. This means that not only does the flat roof change the appearance of the building, but it necessitates the use of different roof framing and roof covering materials. Commercial buildings are designed to be as maintenance free as possible. The wood or vinyl molding that decorates the exterior of many houses would require too much maintenance for a commercial office building. This is not to say that a commercial building cannot be architecturally pleasing. In fact, because more money is generally available to design and construct large commercial buildings, they often tend to be extremely attractive and viewed as works of art. However, the aesthetic qualities in a commercial building are generally built into the overall design of the building, rather than applied after the shell is erected.

Larger buildings require greater strength. The weight of the materials required to build the School Addition is many times the weight of the materials to build even a large house. Also, the external loads (wind, snow, and so forth) on a larger building are greater than those on a smaller building. All these greater loads require stronger structural members, more precisely engineered construction methods, and fastenings that transmit the loads from one structural member to the next. This usually dictates the use of structural steel and reinforced concrete for the structural elements of the building—the frame and foundation.

## Structural Steel

Steel is made by alloying small amounts of carbon and other elements with iron. The amount of carbon in the steel determines its toughness. Other elements, such as chromium, copper, nickel, or titanium, are used to produce specific properties. Stainless steel, for example, is made by alloying chromium

**Figure 37–1.** Generally, medium to large commercial buildings are very functional in design.

with the steel. Structural steel is mild steel (a small amount of carbon is used) that is rolled into specific shapes used for construction. The most common structural shapes are shown in **Figure 37–2.** W shapes are used for most beams and columns in steel-frame buildings. The designator for a structural steel member includes a symbol or capital letter to indicate the basic shape, one or more numbers to indicate the size, and a weight-per-foot designation. For example, drawing S200 for the School Addition, **Figure 37–3,** shows a W6 × 25 column. This is a *wide*-flange shape, with a *6*-inch web, weighing *25* pounds per foot. Standard dimensions for structural steel shapes are published by the American Institute for Steel Construction (AISC) *Manual of Steel Construction.*

Structural steel members are most often joined by welding. Bolts are sometimes used to attach dissimilar materials, such as wood and steel framing members. Where bolts are to be used, they are usually shown by either symbols or a drawing of a bolt on the details, **Figure 37–4.** Detail J-14 of the drawings for the School Addition (see **Figure 37–4**) introduces a technique that is often used on drawings for commercial construction. Instead of describing all the information on the detail drawing itself, many of the drawing callouts refer to a material legend. The *material legend,* which is a running list on the right side of most of the school drawings, identifies all the material items in the building by a legend number.

J-14 shows four items identified only as 5.10D. The material legend on Sheet A504 identifies these as anchor bolts.

## Welding Symbol

Welds are depicted in drawings by the use of standard welding symbols, **Figure 37–5.** The American Welding Society (AWS) has developed a standard system for welding symbols. Each part of the welding symbol has a specific meaning. The arrow indicates what steel parts are to be welded. The term *welding symbol* should not be confused with *weld symbol.* The welding symbol is the complete symbol, including arrow, reference line, and any information added to it. The weld symbol indicates the type of weld to be made. The type of weld used most often in structural steel for construction is called a *fillet weld* and is indicated by a triangle. Other types of welds include plug welds, groove welds, and butt welds, **Figure 37–6.** The reference line is always drawn parallel to the bottom edge of the drawing. The lower side of the reference line is called the *arrow side.* Any information, such as the type of weld, that is shown below the reference line applies to the side of the assembly to which the arrow points. The upper side of the reference line is called the *other side.* Information above the reference line applies to the side of the assembly away from the arrow, **Figure 37–7.** The tail of the arrow is used to specify welding processes, electrode materials, and any

| Shape | Letter Designation | Size Designation |
|---|---|---|
| Wide Flange | W | depth of web × weight per foot |
| Standard Beam | S | depth of web × weight per foot |
| Channel or Miscellaneous Channel | C or MC | depth × weight per foot |
| Structural Tube | TS | width × depth |
| Tee | T | depth × weight per foot |
| Angle | L | vertical leg × horizontal leg × thickness |

**Figure 37–2.** Common structural shapes and their designations.

other specifications that cannot be shown elsewhere on the welding symbol.

## Shop Fabrication

Many structural subassemblies are more easily or more accurately fabricated in a shop environment. For example, if a special beam is to be installed with a welded subassembly on its ends, the subassembly would probably be fabricated in the steel shop so the beam can be delivered to the site ready for erection. A large structural steel frame might require as many *fabrication drawings* (sometimes called *shop drawings*) as structural drawings. Usually, the field contractors do not deal with shop

W6x25 COLUMN

3/4" x 9" x 9" BASE ℄
OVER 1/4" LEVELING ℄
AND 3/4" GROUT PAD

COL.
LINE

COL.
LINE

(2) 3/4" ⌀ A.B.'s

EQ
EQ
EQ

2½"  2½"

EQ  EQ

TYP  1/4

1'-3"

3"

# TYPICAL
# BASE PLATE DETAIL
### (28 LOCATIONS)

3
S200

**Figure 37–3.** Structural steel designation on a drawing.

drawings, and they are not covered in this book. However, it is important to recognize notes on the structural drawings that indicate a difference between field erection and shop fabrication. One example of such information is the field weld symbol, **Figure 37–8.** The field weld symbol is a small flag at the break in the arrow of the welding drawing. This symbol indicates the weld is to be done in the field, not in the fabrication shop.

## Foundations for Commercial Construction

Houses and small frame buildings usually have very simple foundation systems. The superstructure does not impose a heavy load on the foundation, so a simple *haunched slab* or *inverted-T foundation* is all that is required to support the building in most soil conditions. As buildings become larger and more complex, they place a greater load on the foundation system. Also, larger buildings tend to be designed with much of the load being transmitted through a relatively small number of columns, instead of being spread over a sill member by closely spaced studs. To support these loads, the foundations of heavy buildings are carefully engineered by structural engineers and are more heavily reinforced than the foundations for smaller buildings.

Generally, continuous wall foundations are similar to those for smaller buildings. They consist of a *spread footing* that supports a continuous concrete wall. Because of the greater size of the structure, the footing and wall may

SEE
G-3/A520

7.50A
5.10 D
6.10 B(½)
6.10 A

7.50H

4.20C(10")
FIRE HALL

2" MIN. ½" ½" 2¾" 5¼"

4.20C(10")
BOND BEAM

1'-7½"

5.10 D
5.10 B (7"×4"×⅜")

| ROOF EDGE DETAIL | J-14 |
|---|---|
| A201 | 3"=1'-0' |

**Figure 37–4.** Bolts are sometimes used for fastening structural members.

TAIL, USED TO
SHOW WELDING
PROCESS OR OTHER
SPECIFICATIONS

WELD SYMBOL
SHOWS TYPE AND
LOCATION OF WELD

REFERENCE LINE
(ALWAYS PARALLEL
TO THE BOTTOM
OF THE DRAWING)

ARROW INDICATES
PARTS TO BE WELDED

**Figure 37–5.** The basic welding symbol.

| FILLET | PLUG OR SLOT | SPOT OR PROJECTION | STUD | SEAM | BACK OR BACKING | SURFACING | SCARF (FOR BRAZED JOINT) |
|---|---|---|---|---|---|---|---|
| | | | | | | | |
| | | | | | | | |

| FLANGE | | GROOVE | | | | | | |
|---|---|---|---|---|---|---|---|---|
| EDGE | CORNER | SQUARE | V | BEVEL | U | J | FLARE-V | FLARE-BEVEL |
| | | | | | | | | |
| | | | | | | | | |

NOTE: _ _ _ _ REPRESENTS REFERENCE LINE WHICH IS NOT SHOWN AS A SOLID LINE FOR THIS PURPOSE.

**Figure 37–6.** Weld symbols used to show various types of welds.

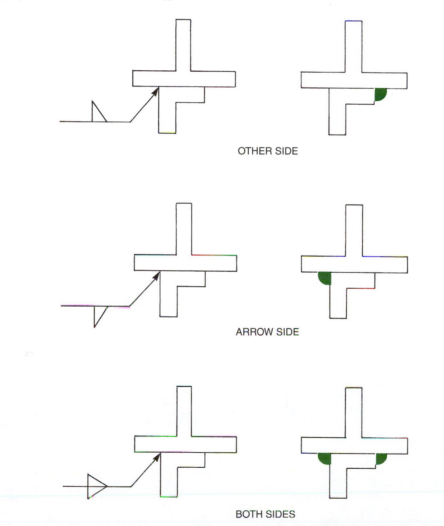

OTHER SIDE

ARROW SIDE

BOTH SIDES

**Figure 37–7.** Arrow-side, other-side significance.

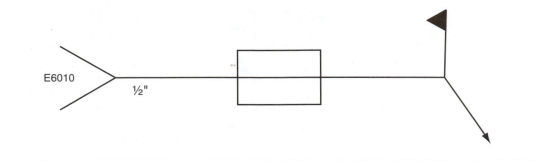

**Figure 37–8.** This can be recognized as a field weld symbol by the flag. It is a ½-inch plug weld on both sides of the part and is to be done with E6010 electrode or filler metal.

be more massive for a heavy building and will probably require more and larger-diameter reinforcing steel. The overall layout of the foundation is shown on a foundation plan, **Figure 37–9.** The dimensions and the reinforcement of the footings and walls are planned by engineers to carry the necessary loads, so it is likely that there will be several detail drawings to further describe the foundation systems at different points in the building, **Figure 37–10.** Some of the reinforcement may be bars that protrude only a short distance out of the footing and into the wall. Reinforcing bars used in this way are called *dowels* and are used to keep the wall from shifting on the footing.

Column footings are located by the column centerlines. All these footings must be centered under the columns they support unless specifically noted otherwise. The outline of the footing is shown on the foundation plan using dashed lines. This indicates that the footing is actually hidden from view by the concrete slab. Column footing dimensions are usually shown in a schedule, **Figure 37–11.** There are usually several column footings of the same size. The mark-schedule system allows the drafter to show all the similar footings with a single entry on the schedule.

*Piles* are long poles or steel members that are driven into the earth to support columns or other grade beams. Pile foundations support their loads by means of several piles in a cluster, topped with a pile cap made of reinforced concrete. Pile foundations are not as common as spread footings and continuous walls. *Grade beams* are reinforced concrete beams placed at grade (ground) level to provide a bearing surface for the superstructure.

**Figure 37–9.** The layout of the foundation is shown on the foundation plan.

---

## ✓ CHECK YOUR PROGRESS

Can you perform these tasks?

● Explain several differences between typical light-frame construction and larger commercial construction.

● Describe common structural shapes from their identification on drawings.

● Explain how the foundations for large buildings are different from the foundations for single-family homes.

6" REINFORCED CMU WALL
SEE 1/203 FOR REINFORCING
REQUIREMENTS

1/2" EXPANSION
JOINT MATERIAL (TYP)

5" SLAB ON GRADE

VAPOR BARRIER

6" COMPACTED
SUBBASE MATERIAL
2" RIGID INSUL
(SEE ARCH.)

#4 DOWELS @ 16" O.C.
ROUGHEN JOINT

(2)-#5 CONT BARS

UNDISTURBED SOIL OR
COMPACTED STRUCTURAL
FILL (TYP)

COL LINE

12 1/2"

3"

SEE ARCH. DWGS
FOR FLASHING &
ADDITIONAL
INFO. (TYP.)

HORIZ. JOINT REINF
EVERY OTHER COURSE

4" BRICK VENEER

ELEV. = 99'-2"

FINISHED GRADE
(SEE SITE DWGS)

#4 VERT. BARS @
16" O.C. E.F.

#4 HORIZ BARS @
12" O.C. E.F.

SEE PLAN FOR
FTG. ELEV (TYP)

1/2"

1 1/2"
COVER

1'-0"

1'-4" LAP

3" CLR.

1'-0"

2'-0"

## 13/S101 TYPICAL WALL SECTION
SCALE: 3/4"=1'-0"

**Figure 37–10.** Foundation detail.

## FOOTING SCHEDULE

| MARK | SIZE | REINF. | REMARKS |
|------|------|--------|---------|
| F1 | 3'-0"x 3'-0"x 12" | (4)#5 BARS EA. WAY | |
| F2 | 4'-0"x 4'-0"x 12" | (5)#5 BARS EA. WAY | |
| F3 | 4'-6"x 4'-6"x 12" | (5)#6 BARS EA. WAY | |
| F4 | 3'-0"x 4'-6"x 24" | (5)#5 SHORT BARS (7)#5 LONG BARS | OVERPOUR EXISTING BUILDING FOOTING (SIM. TO 14/S101) |
| | | | |
| | | | |
| | | | |

**Figure 37–11.** Footing schedule.

## ASSIGNMENT

1. List five features of the School Addition that would be very different on a light-frame building, and explain the differences.

2. Describe each of the structural steel designations listed below:
   W14 × 82
   MC12 × 45
   L6 × 3½ × ⁵⁄₁₆

3. What is the structural steel shape and depth of the steel beam shown supporting the roof bar joists in detail 3/S202?

4. Draw a sketch showing where the welds would be made for each of the following symbols:

5. Describe the information given in the welding symbol on detail 3/S202.

6. Based on the information shown on Sheet S100, would you say that the School Addition has a pier foundation, grade beams, or a continuous wall foundation with spread footings?

7. How many pieces of reinforcing steel are shown in the footing on Wall Section 13/S101? Describe each, including its size and purpose.

8. How many 4'−6" × 4'−6" × 12" column footings are shown on the foundation plan for the School Addition?

9. There are several pairs of dashed (hidden) lines running vertically through the interior of the foundation plan. A note on one of these pairs of lines indicates that the slab is to be thickened. Why is it thickened at these points? How much thicker is it at these points than the surrounding area? What additional reinforcement is provided in the thickened portion of the slab?

# UNIT 38

# Coordination of Drawings

## Objectives

After completing this unit, you will be able to perform the following tasks:

○ Interpret the indexes and keys normally found on drawings for commercial construction.

○ Use the number system for referencing details, sections, and other drawings.

○ Explain the use of a grid system to locate columns and piers.

○ Use the plans, elevations, and details to describe the layout of the rooms and spaces in a large commercial building.

## Indexes

A set of working drawings for a large commercial building might fill more than one hundred sheets. The School Addition, although it is small by comparison with many large buildings, is part of a renovation and expansion program that spans three buildings and is described on sixty-six drawing sheets. A key plan in the lower right corners of Sheets A101 and A102 shows the addition in relationship to the rest of the Junior/Senior High School building. Finding one's way around these sixty-six sheets would be a difficult task without some kind of an index. As with most commercial construction drawing sets, the cover page of the drawing set includes the index, **Figure 38–1.**

It is not reasonable to include all sixty-six drawing sheets, plus the cover page, with this textbook because that would add the equivalent of more than two hundred pages to the book. Therefore, the drawing packet with this text includes all the drawings that are necessary to completely understand the additions to the Junior/Senior High School but that were not reprinted on the pages of the text. The Index of Drawings, found in **Figure 38–1,** lists all the original drawing numbers. The drawing numbers that are underlined indicate the drawings contained in the drawing packet. Parts of other sheets are included as illustrations in this textbook. Drawings for civil (site work) that are not needed for an understanding of the structure and drawings for work on other school property are not included.

The *drawing index* includes the number of the drawing and the drawing title. *Drawing numbers* are made up of one or two letters and numerals. The letters indicate the basic type of construction shown on the drawing. Most architects use the letter *A* to indicate architectural drawings, *S* to indicate structural drawings, *C* for civil, and so on. The numbers indicate the sequence of the drawings in that section. Some architects indicate levels of drawings by using a series within a range to designate a level. For example, 100 through 199 might be reserved for foundations, 200 through 299 might be floor plans and elevations showing the overall superstructure, and 300 through 399 might be roof framing. This system is usually quite easy to spot by studying the index briefly.

## Material Keying

In light-frame construction of small buildings, it is customary for callouts on the drawings to label the material used. On larger projects, however, it is customary for a material keying system to be used so that the drawings themselves are

| | GENERAL CONSTRUCTION CONTRACT |
|---|---|
| A000 | MASTER KEYNOTES |
| G001 | SITE PLAN (JR/SR HIGH SCHOOL) |
| G002 | KEY PLANS (JR/SR HIGH SCHOOL) |
| G003 | KEY PLANS (BERNE ELEMENTARY & WESTERLO ELEMENTARY |
| C100 | SITE DEMOLITION PLAN (JR/SR HIGH SCHOOL & BERNE ELEMENTARY |
| C101 | LAYOUT PLAN (JR/SR HIGH SCHOOL) |
| C102 | GRADING PLAN (JR/SR HIGH SCHOOL) |
| C103 | PLANS & DETAILS (JR/SR HIGH SCHOOL) |
| C201 | PARTIAL SITE PLAN, SOUTH (JR/SR HIGH SCHOOL) |
| C202 | PARTIAL SITE PLAN, NORTH (JR/SR HIGH SCHOOL) |
| C203 | PLAN AND DETAILS - SEWAGE DISPOSAL (JR/SR HIGH SCHOOL & BERNE ELEMENTARY) |
| C204 | PLAN AND DETAILS; NEW WELL (JR/SR HIGH SCHOOL & BERNE ELEMENTARY) |
| C205 | DRAINAGE PROFILES AND DETAILS (JR/SR HIGH SCHOOL & BERNE ELEMENTARY) |
| C206 | DETAILS SEWAGE DISPOSAL (JR/SR HIGH SCHOOL & BERNE ELEMENTARY) |
| C207 | PLANS AND DETAILS (JR/SR HIGH SCHOOL & BERNE ELEMENTARY) |
| C208 | SECTIONS AND DETAILS (JR/SR HIGH SCHOOL & BERNE ELEMENTARY) |
| A001 | ROOM FINISH SCHEDULE, DOOR SCHEDULE AND DETAILS (JR/SR HIGH SCHOOL) |
| A101 | FIRST FLOOR PLAN (JR/SR HIGH SCHOOL) |
| A102 | SECOND FLOOR PLAN (JR/SR HIGH SCHOOL) |
| A103 | CAFETERIA & BOILER ROOM PLAN, DEMOLITION PLAN & DETAILS (JR/SR HIGH SCHOOL) |
| A104 | FLOOR PLAN & DETAILS (BUS GARAGE) |
| A201 | EXTERIOR ELEVATIONS & BUILDING SECTIONS (JR/SR HIGH SCHOOL) |
| A301 | STAIR PLANS & SECTIONS (JR/SR HIGH SCHOOL) |
| A302 | STAIR DETAILS (JR/SR HIGH SCHOOL) |
| A303 | RAMP & ELEVATOR (JR/SR HIGH SCHOOL) |
| A501 | WALL SECTIONS (JR/SR HIGH SCHOOL) |
| A502 | WALL SECTIONS (JR/SR HIGH SCHOOL) |
| A503 | PLAN DETAILS (JR/SR HIGH SCHOOL) |
| A504 | WALL DETAILS (JR/SR HIGH SCHOOL) |

**Figure 38–1.** Index from the cover page of school drawings.

| | |
|---|---|
| A510 | WINDOW TYPES AND DETAILS (JR/SR HIGH SCHOOL) |
| A511 | EXTERIOR ELEVATIONS (WESTERLO ELEMENTARY) |
| A512 | WINDOW TYPES & DETAILS (WESTERLO ELEMENTARY) |
| A520 | ROOF PLAN & DETAILS (JR/SR HIGH SCHOOL) |
| A610 | INTERIOR ELEVATIONS & TOILET ROOM FLOOR PLAN (JR/SR HIGH SCHOOL) |
| S100 | FOUNDATION PLAN & NOTES (JR/SR HIGH SCHOOL) |
| S101 | FOUNDATION DETAILS (JR/SR HIGH SCHOOL) |
| S200 | SECOND FLOOR FRAMING PLAN & DETAILS (JR/SR HIGH SCHOOL) |
| S201 | ROOF FRAMING PLAN & DETAILS (JR/SR HIGH SCHOOL) |
| S202 | STEEL FRAMING DETAILS (JR/SR HIGH SCHOOL) |
| S203 | MISCELLANEOUS STEEL DETAILS & MASONRY REINFORCING DETAILS (JR/SR HIGH SCHOOL) |
| | |
| | ROOF RECONSTRUCTION CONTRACT |
| A521 | ROOF PLAN & DETAILS (JR/SR HIGH SCHOOL) |
| A522 | ROOF PLAN & DETAILS (BERNE ELEMENTARY) |
| A523 | ROOF PLAN & DETAILS (BERNE ELEMENTARY) |
| | |
| | ASBESTOS ABATEMENT CONTRACT |
| ASB-1 | ASBESTOS ABATEMENT - PLANS AND DETAILS (JR/SR HIGH SCHOOL) |
| ASB-2 | PARTIAL PLAN - BOILER ROOM (JR/SR HIGH SCHOOL) |
| ASB-201 | ASBESTOS ABATEMENT - PLANS AND DETAIL (WESTERLO ELEMENTARY) |
| | |
| | PLUMBING CONTRACT |
| P-1 | PARTIAL PLAN - ADDITION, FIRST FLOOR AND SECOND FLOOR (JR/SR HIGH SCHOOL) |
| P-2 | PARTIAL PLAN - CAFETERIA EXPANSION; DETAILS (JR/SR HIGH SCHOOL) |
| | |
| | HEATING, VENTILATING & AIR CONDITIONING CONTRACT |
| H-1 | FIRST & SECOND FLOOR PLAN ADDITION; SYMBOL LEGEND (JR/SR HIGH SCHOOL) |
| H-2 | PARTIAL PLAN BOILER ROOM; REMOVALS AND NEW WORK (JR/SR HIGH SCHOOL) |
| H-3 | PARTIAL PLAN - CAFETERIA EXPANSION; DETAILS (JR/SR HIGH SCHOOL) |
| H-4 | SCHEDULES AND DETAILS (JR/SR HIGH SCHOOL) |

**Figure 38–1.** (continued)

| | |
|---|---|
| H-5 | DETAILS (JR/SR HIGH SCHOOL) |
| H101 | PARTIAL PLAN - BOILER ROOM; DETAILS; SCHEDULES (WESTERLO ELEMENTARY) |
| H201 | PARTIAL PLAN, BOILER ROOM, DETAILS, SCHEDULES (WESTERLO ELEMENTARY) |
| H202 | DETAILS AND SCHEDULES (WESTERLO ELEMENTARY) |
| | ELECTRICAL CONTRACT |
| E-1 | FIRST & SECOND FLOOR PLAN, ADDITION LIGHTING (JR/SR HIGH SCHOOL) |
| E-2 | FIRST & SECOND FLOOR PLAN, ADDITION UTILITIES (JR/SR HIGH SCHOOL) |
| E-3 | PARTIAL PLANS CAFETERIA & COMPUTER LABS (JR/SR HIGH SCHOOL) |
| E-4 | BASEMENT PLAN, PART PLANS BOILER ROOM, DETAILS (JR/SR HIGH SCHOOL) |
| E-5 | FIRST FLOOR PLAN, DETAILS (JR/SR HIGH SCHOOL) |
| E-6 | SECOND FLOOR PLAN, DETAILS (JR/SR HIGH SCHOOL) |
| E-7 | SCHEDULES (JR/SR HIGH SCHOOL) |
| E101 | GROUND FLOOR PLAN; DETAILS; SCHEDULES (WESTERLO ELEMENTARY) |
| E102 | FIRST FLOOR PLAN; DETAILS; SCHEDULES (WESTERLO ELEMENTARY) |
| E103 | SECOND FLOOR PLAN; DETAILS; SCHEDULES (WESTERLO ELEMENTARY) |

**Figure 38–1.** (continued)

not cluttered by excessive notation. The following, which is reprinted from the cover page of the School Addition drawings, explains the system:

### MATERIAL KEYING SYSTEM

A KEYNOTING SYSTEM IS USED ON THE DRAWINGS FOR MATERIALS REFERENCES AND NOTES. REFER TO THE KEYNOTE LEGEND ON THE DRAWINGS FOR THE INFORMATION WHICH RELATES TO EACH KEYNOTE SYMBOL ON THE RESPECTIVE DRAWING.

EACH KEYNOTE SYMBOL CONSISTS OF A NUMBER FOLLOWED BY A LETTER SUFFIX. THE NUMBER RELATES TO THE SPECIFICATION SECTION WHICH GENERALLY COVERS THE ITEM THAT IS REFERENCED, AND THE LETTER SUFFIX COMBINED WITH THE NUMBER CREATES A KEYNOTE SYMBOL WHICH IDENTIFIES THE SPECIFIC REFERENCE NOTATION USED ON THE DRAWING. THE LETTER SUFFIX DOES NOT RELATE TO ANY CORRESPONDING REFERENCE LETTER IN THE SPECIFICATION.

THE ORGANIZATION OF THE KEYNOTING SYSTEM ON THE DRAWINGS, WITH THE KEYNOTE REFERENCE NUMBERS RELATED TO THE SPECIFICATIONS SECTIONS NUMBERING SYSTEM, SHALL NOT CONTROL THE CONTRACTOR IN DIVIDING THE WORK AMONG SUBCONTRACTORS OR IN ESTABLISHING THE EXTENT OF WORK TO BE PERFORMED BY ANY TRADE.

**Figure 38–2** illustrates how this keynote system works. Notice that many of the callouts on this detail are a decimal number followed by a letter. That is the keynote symbol. Referring to the Index of Drawings, **Figure 38–1**, we find that Sheet A000 is Master Keynotes. The complete legend of Master Keynotes, as it appeared on Sheet A000, is reprinted on pages 336 through 343. The keynotes that pertain to a particular drawing sheet are also copied on the right end of the drawing sheet. In **Figure 38–2** the keynotes on the right side of the figure pertain to the drawings that were together on Sheet A302, where J-14

**Figure 38–2.** Material keynote system.

was located. The top right callout on J-14 has a keynote symbol, 3.30A. Referring to the material keying legend, 3.30A is cast-in-place concrete. This keynote replaces the customary callout on the drawing that would indicate that the landing and tread surfaces are cast-in-place concrete. More specific information, such as the grade of concrete and type of finish, would be indicated in Section 3 of the specifications. Moving counterclockwise around the drawing, the next keynote symbol is 5.50E and is referenced to the stair stringer that is *beyond* the main part of the drawing. According to the material keying legend, 5.50E is covered in Section 5.50 of the specifications, Miscellaneous Metals, and is an MC12 × 10.6 stringer. (Unit 39 explains the meaning of this structural steel designation.)

Many of the material keynotes are referenced on several of the drawing sheets. In that case, they are reproduced on each of those sheets, as well as on the master legend on Sheet A000. All of the whole sheets with this book include the material keying legends as they were intended by the architects.

## Reference Symbols Used on Drawings

A large building, like the School Addition, requires a large number of sections and details to describe all aspects of the structure. The symbols for section views and the symbols for detail drawings are similar in appearance and format but have one significant difference. While both commonly use a circle as a symbol, the circle on

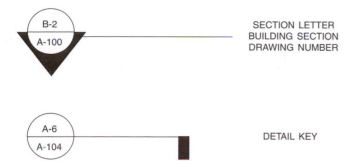

SECTION LETTER
BUILDING SECTION
DRAWING NUMBER

**Figure 38–3.** Detail and section-view symbols.

INTERIOR ELEVATION

L-10/A103

**Figure 38–4.** Elevation reference symbol.

DETAIL KEY

a plan or other large-scale drawing to indicate existing detail contains the detail number above a horizontal line. Below the horizontal line is the sheet number on which the detail appears. Some element of the symbol, such as a rectangular flag, indicates that it is a detail drawing symbol. Section view symbols use the same scheme for specifying section number and sheet number but use a pointer instead of a rectangular flag to indicate the direction in which the section is viewed. **Figure 38–3** is reprinted from the cover page for the School Addition. In this case, Section B-2, which would be found on Sheet A100, is drawn with a view of the building from the top of the page toward the bottom. (The Index of Drawings shows that the A series of drawings includes A000 and A101 through A610, but no A100. Apparently, A100 is just a hypothetical sheet for the sake of explaining how symbols are used.)

An example from the drawings that are used with this textbook can be seen on the First Floor Plan, Sheet A101. At the extreme right and left ends of the floor plan there is a section-view symbol, with its pointers directed toward the top of the drawing (north, according to the symbol in the lower right corner of the drawing). This symbol refers to Section View J-14 on Sheet A201. Section J-14 is in the top right of Sheet A201 and is a full cross-section view of the building as seen by a person standing on the south side of the building and looking north. The detail drawing symbol also can be found on the First Floor Plan, located just above the section-view symbol at the left end of the building. The direction of the rectangular flag indicates that it is also drawn as though viewed while facing north. That detail, which is at a larger scale (¾″ = 1′−0″) than the building sections (⅛″ = 1′−0″), is number A-4 on Sheet A502. Detail A-4 includes further detail symbols, such as one about halfway up the wall referring to E-2 on A504, an expansion joint detail, at 3″ = 1′−0″.

Interior elevations, which are often used to show the locations of fixtures, built-in furniture, and special architectural details, are usually referenced with their own type of symbols on floor plans. The symbol for an interior elevation also shows the direction from which the elevation is viewed, and it includes the number of the elevation and the sheet on which that elevation drawing appears. The symbol in **Figure 38–4** references elevation drawing L-10, which would be found on Sheet A103. The direction the triangular shape of the symbol points is the direction in which the elevation is viewed.

## Structural Grid Coordinate

Buildings with columns of reinforced concrete or structural steel as a main element of the building frame are laid out on a structural grid. The plans are drawn with a system of horizontal and vertical reference lines, so that all columns and details can be referenced to this grid. Vertical lines are numbered, and horizontal lines are lettered; or the system is reversed, with horizontal lines numbered and vertical lines lettered, **Figure 38–5**. Columns are located by grid lines through their centers. The locations of major architectural components, such as walls, are located by dimensions that are referenced to the grid lines. The dimensions from the grid lines to the faces of exterior walls are usually given at the corners of the building. Interior partitions and other architectural elements may be located by dimensioning to the faces of these exterior walls, which are referenced to the structural grid.

## Mental Walk-Through

As with each of the buildings discussed earlier in this textbook, one of the first steps you should take to familiarize yourself with the School Addition is to take a *mental walk-through*. Start at the front entrance to the first floor,

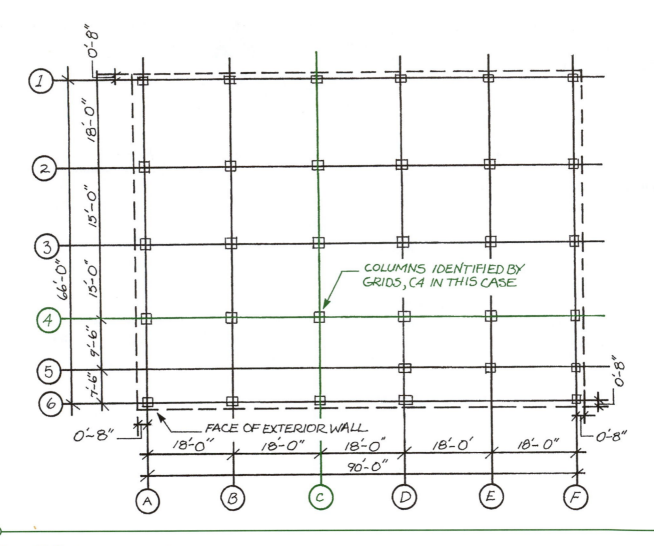

**Figure 38–5.** Structural grid coordinate system.

in the lower right corner of the First Floor Plan. Enter door 101, and step into a stairwell marked stair #1. At the far end of the stairwell is door 102, which opens into a corridor. If you turn right in the corridor, you can walk up a ramp into the existing building, 10 inches higher than the first floor of the addition. Across the main corridor of the addition are the doors into two toilet rooms. A symbol in front of those doors indicates that Elevation A-8 on A610 will show you what the toilet room doors look like. As you continue your walk-through, you will come to many more elevation symbols. Look at each of the elevations on your walk-through to help you form a mental picture of the addition. You will also see several detail and section symbols. It will not be necessary to check each of these to form a mental picture of the addition, but you should look up a couple of each, just to make sure you are comfortable locating them.

Moving to the west in the main corridor, you come to a door (103) on your left. This is the entrance to a storage room. A dimension inside the storage room indicates that it is 13′−8½″ wide. The exterior wall (south wall) of the storage room is on grid line F (see the left end of the floor plan). The wall with the door is on grid line D, which is 32′−7½″ from grid line F. You could determine the actual face-to-face dimension between the north and south walls by looking at the details and sections, but that is not necessary for this mental picture of the building. Notice that where the east and west walls of the storage room intersect with the wall of the corridor, there are two heavy dashed crosses, indicating columns.

Most walls are on column lines. These are on vertical grid lines 6 and 7.

Now walk through the janitor closet (JAN. CL.), each of the classrooms, and all corridors and stairwells. Only a few highlights are described in this textbook, but after you have followed the walk-through described here, you should take a second walk-through, noticing and studying everything until you can visualize it.

Along both sides of the corridor are several references to Material Keynote 10.50A. Find that keynote on the legend at the right end of the drawing sheet or on the master keynote legend.

Through door 105 you enter classroom 103. On the floor plan, room 103 is only separated from room 104 by a single dashed line. That warrants investigation. A detail symbol directs you to A-6 on Sheet A504. That is a large-scale section view of a track detail. Looking again at the floor plan, you see a material keynote reference near the lower end of the dashed line: 10.62A. According to the legend, 10.62A is a folding partition, so rooms 103 and 104 are separated by a folding partition.

Across the corridor from classroom 103 is stair #2, which goes to the second floor. At the extreme left end of the corridor is a large rectangular area identified by keynote 14.20A, an elevator. Take the elevator to the second floor.

When you step off the elevator, you are on the second floor. Your walk-through is continued on Sheet A102, Second Floor Plan. The second floor of the addition is almost identical to the first floor. It would be foolhardy to take for granted that the two levels are identical, but a quick survey shows that they have the same rooms in approximately the same layout. The rooms on the second floor are numbered in the 200 series to indicate that they are on the second floor.

## ✓ CHECK YOUR PROGRESS

Can you perform these tasks?

- Find and understand the information given in the Index to Drawings and the materials specifications on drawings.
- Cross-reference between sections, details, and other drawing sheets on the basis of the numbering system on the drawings.
- Locate columns according to their grid coordinates.
- Find your way around plans and elevations well enough to form a mental picture of the building.

## ASSIGNMENT

Refer to the drawings of the School Addition (included in the packet) to complete this assignment.

1. List the sheet numbers for all the structural drawings for the addition.

2. On what sheet would you look to find wall details?

3. On what sheet would you find a framing plan for the roof?

4. What is indicated by Keynote 10.80C?

5. What keynote number is used to indicate plywood?

6. On the First Floor Plan, at about the midpoint of the north wall, a circle is drawn over a triangular flag. The notation inside the circle is J-6/A201. Explain what this symbol indicates and where further information can be found.

7. According to the building elevations and several wall details, there is a precast concrete sill or band that is about waist high on the exterior walls. What is the dimension from the top of the foundation to the bottom of this sill or band?

8. What is the height of the concrete block wall on the north wall of the first floor corridor?

9. What is the material on the lower surface of the canopy over the entrance at door 101?

10. What are the dimensions of the liquid marker board and tackboards on the west wall of classroom 107?

11. What are the dimensions of the mirrors in the toilet rooms?

12. Describe the location of column D3, relative to the floor plans.

13. What is the center-to-center spacing between columns F7 and F9?

14. What are the overall outside dimensions of the pre-cast concrete band, such as is used near the base of typical exterior walls?

15. There is a door from a classroom (112) in the existing building into the storage room in the northeast corner of the addition. The door is made of what material?

16. Describe the differences between the rooms adjacent to the elevator on the first floor and on the second floor.

17. Where do you find the reflected ceiling plan?

18. Is the electrical service entrance section inside the existing building or outside the existing building?

19. How many light switches with motion sensors are there? Where are they located?

20. Which lighting fixtures are scheduled to have electronic ballasts?

# Structural Drawings

## Foundations for Commercial Buildings

The foundations for large, commercial buildings perform the same functions as those for light-frame buildings. The foundation supports the loads (weight) imposed on the superstructure and spreads those loads over a large enough area so that the earth can support it uniformly. In most commercial construction the foundation system is composed of spread footings and stem walls and pads that act as footings under columns. These are the same elements that are found in most one- and two-family homes. The biggest difference in the foundation for a commercial building and that for a small house is the thickness of the concrete and the amount of reinforcing steel. There are apt to be more detail drawings for the foundation of a larger building, because of the need to describe the different sizes and shapes of footings and the reinforcement at many points in the foundation. The same knowledge is required to read the foundation plan for a large building as for a smaller one.

The drawings consist of a plan that shows the layout of the foundation and the major dimensions and detail drawings that describe reinforcement, expansion joints, and variations in design at special locations. The placement of column footings is indicated by referencing the structural grid coordinates. The structural grid indicates where the centerline of the column will be, but it is very important that the footing be placed accurately on this center. If the column does not rest on the center of the footing, the footing may tip due to the uneven pressure on the soil below, **Figure 39–1.**

## Structural Steel Framing

A structural steel building frame consists of **columns** (vertical members) and *beams* (horizontal members.) The largest beams, called *girders,* attach to the columns, **Figure 39–2.** *Joists* are intermediate beams and are supported at their ends by the girders. *Lintels* are the beams that support the weight above an opening, such as a door, window, or nonstructural panel. If the first floor is a concrete slab, it is described on the foundation plans and details. The second and higher floors are described on framing plans, as is the roof frame. Many buildings have several floors that are framed alike. In this case a note might indicate that a framing plan is typical.

## Objectives

After completing this unit, you will be able to perform the following tasks:

O Describe the footings for columns and walls, including dimensions and reinforcement.

O Interpret the information found on a foundation plan, including dimensions of foundation walls, reinforcement of foundations, and the locations of the various elements.

O Describe each column, beam, and lintel shown on the structural drawings.

O Interpret structural details and sections.

PRESSURE EVEN
COL. CENTERED

COL. OFF CENTER

FOOTING TIPPED

**Figure 39–1.** The column load must be centered on the footing.

JOISTS

GIRDERS

LINTELS

COLUMNS

**Figure 39–2.** Major parts of a building frame.

Framing plans for structural steel are drawn on the structural grid coordinate system, **Figure 39–3.** Beams are shown by a single-line symbol with an accompanying note to indicate the structural shape to be used. The lines indicating beams stop short of the symbol for the girder or column when the beam is framed into the supporting member and does not continue over it. The abbreviation *do,* which stands for *ditto,* indicates that the specification for the first member in a series is to be repeated for all members. A number in parentheses at the end of the designation is used to indicate the elevation of the top of the member. This may be the elevation or the distance above or below the floor line, **Figure 39–4.**

Joists are frequently *open-web steel joists,* sometimes called *bar joists,* **Figure 39–5.** Bar joists are manufactured in H, J, and K series, depending on the grade of steel used and their strength requirements. K series are stronger than H series, and H series are stronger than J series. A bar joist designation includes a number to indicate depth, a letter to indicate strength series, and a number to indicate chord size. For example, the designation 16H6 indicates a 16-inch-deep H series joist with number 6 chords.

The actual lengths of members are not shown on the general contract drawings. This information is shown on shop drawings that are drawn by the steel fabricator some

**Figure 39–3.** Structural coordinates are shown on the framing plans and floor plans.

EL. FIN. FLOOR 74'-4"
TOP OF STEEL ABOVE OR
BELOW FIN. FL. AS NOTED
THUS (+  ) OR (–  )

FIN..FL. EL. 74'-4"
EL. TOP OF STEEL BEAMS
SHOWN THUS (  )

**Figure 39–4.** Notations on the drawings indicate the relative elevations of beams.

TOP CHORD

BOTTOM CHORD

OPEN WEB-USUALLY
MADE OF ROUND RODS

SUPPORT POINT NEAR TOP OF JOIST

**Figure 39–5.** Open-web steel joists (often called bar joists). *Courtesy of Willis: Blueprint Reading for Commercial Construction, by Delmar Learning.*

time after the construction drawings are completed. It is an easy matter, however, to find the span of the member by looking at the framing plan. Connections are shown on details and sections.

Lintels are sometimes categorized as *loose steel.* In masonry construction, a steel lintel is placed above each opening to support the weight of the masonry above the opening, **Figure 39–6.** These lintels are not attached to the steel building frame, which is why they are called loose steel. If there are several lintels alike, they are normally shown on the plans by a symbol and are then described more fully in a lintel schedule, **Figure 39–7.**

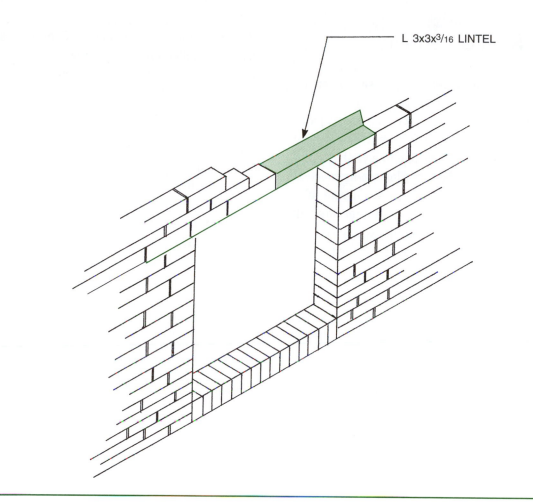

L 3x3x³/16 LINTEL

Figure 39–6. A lintel may be considered loose steel if it is not attached to the frame.

## LINTEL SCHEDULE

| MARK | MATERIAL | TYPE | MAS. OPNG | REMARKS |
|---|---|---|---|---|
| L1 | WT8×13 | ⊥ | SEE ARCH. | ATTACH ONE END TO COL. SEE 1/S200 |
| L2 | WT8×13 | ⊥ | " | END BEARING BOTH ENDS |
| L3 | WT8×13 | ⊥ | " | CONTINUOUS W/CENTER SUPPORT ON T.S. 4×4 |
| L4 | WT4×9 | ⊥ | " | AT ALL BELOW WINDOW UNIT VENTILATOR OPENINGS |
| L5 | (2) ∟ 6×4×5/16 W/ ¼"×9½ CONT. BOT. PL | ⊥⊥ | " | LINTEL IN FIRE WALL WELD BOT. PL TO ANGLES. |
| L6 | (2) ∟ 6×4×5/16 | ⊥⊥ | " | IN EXIST. BLDG. WALL |
| L7 | (3) ∟ 4×3½×5/16 | ⊥⊥⊥ | " | AT CAFETERIA HVAC UNIT WALL OPENINGS COORDINATE WITH G-7/H3 AND A103 (2-LOCATIONS) |

Figure 39–7. Lintel schedule.

# Masonry Reinforcement

Masonry materials have great compressive strength. That is, they resist crushing quite well, but they do not have good tensile strength. The **mortar** in masonry joints is especially poor at resisting the forces that tend to pull it apart, such as a force against the side of a wall or a tendency for the wall to topple. Masonry joint reinforcement is done by embedding specially made welded-wire reinforcement in the joints, **Figure 39–8.** Greater strength can be achieved by building the masonry wall with reinforcement bars in the cores of the masonry units, then filling those cores with concrete. Concrete used for this purpose is called *grout.* Reinforcing steel is also embedded in masonry *walls* to tie structural elements together. For example, it is quite common for rebars to protrude out of the foundation and into the exterior masonry walls. The strength of a masonry wall can be increased considerably by the use of bond beams, **Figure 39–9.** A *bond beam* is made by placing a course of U-shaped masonry units at the top of the wall. Reinforcing steel is placed in the channel formed by the U shape; then the channel is filled with grout. The result is a reinforced beam at the top of the wall.

REINFORCEMENT EMBEDDED IN MORTAR JOINT

WELDS

WELDED WIRE REINFORCEMENT

**Figure 39–8.** Masonry joint reinforcement.

1 OR 2 REBARS IMBEDDED IN GROUT

PORTLAND CEMENT GROUT

U-SHAPED LINTEL BLOCKS

**Figure 39–9.** A bond beam is used to strengthen the top of the wall.

## ✓ CHECK YOUR PROGRESS

Can you perform these tasks?

● Using the information on drawings, describe the footings for a building.

● Using the information on drawings, give complete information about the foundation for a building.

● Describe the columns, beams, and lintels shown on drawings.

## ASSIGNMENT

Refer to the drawings of the School Addition (included in the packet) to complete this assignment.

1. What are the dimensions of the footing for the northwestern most column?

2. What is the width and depth of the footing at the west end of the addition?

3. What is the elevation of the top of the floor in the elevator pit?

4. How many lineal feet of #5 reinforcement bars are needed for the footing under the east wall of the addition?

5. What size and kind of material is used to prevent the foundation wall from moving on the footings? How much of this material is needed for the west end of the addition?

6. How many pieces of what size reinforcing steel are to be used in the footing for the column between the entrances to classrooms 103 and 104?

7. How closely is the vertical reinforcement spaced in a typical section of the foundation wall?

8. What is the overall length of each piece of rebar used to secure the interior masonry partitions to the concrete slab?

9. What is the spacing of the dowels used to secure the exterior masonry walls to the foundation?

10. Describe how the corridor walls are secured to the columns.

11. How many pieces of W18 × 50 steel are used in the construction of the addition?

12. What size and shape structural steel supports the north ends of the floor joists under classroom 203?

13. What size and shape structural steel supports the ends of the girder in the second floor at the front (side with the door) of the elevator shaft?

14. What is the nominal depth of the joist in the second floor corridor?

15. What is the elevation of the top of the second floor girder between D5 and D6, relative to the second finish floor elevation of 110'-3"?

16. What is the shape and size of the member that supports the masonry above the windows in classroom 108?

# Mechanical Drawings

## Heating, Ventilating, and Air Conditioning Plans

The heating, ventilating, and air-conditioning system for a building larger than a single-family house is designed by an engineer who specializes in this work. The drawings for this work are usually drawn on a basic outline of the floor plan for the building. This allows the mechanical drawings to show where HVAC and plumbing equipment is located in relation to rooms, walls, floors, and so on. The coordination of structural, mechanical, and electrical work is often an issue of some concern on construction projects, and so it is important to understand where and how pipes will pass between beams or through walls. For the sake of this book, we are limiting the mechanical drawings to those for HVAC and plumbing, and we will look at the HVAC drawings first.

Most commercial buildings are heated by burning fuel (natural gas or fuel oil) and distributing the resulting heat in the form of warm air, hot water, or steam. Warm-air heating systems require ducts throughout the building to carry the warm air to the occupied spaces and to return the cool air to the source. A warm-air system has the advantage of allowing for inexpensive air conditioning. The same ducts can be used to distribute either warm air in winter or cool air in summer. The ducts, however, require much more space in ceilings and walls and are harder to route through and between structural elements than are pipes. When hot water or steam is used to transport the heat to the occupied space, two pipes are required: one to supply the hot water or steam and one to return the cooler medium to its source. The basic type of system is readily apparent to anyone looking at the mechanical plans. The plans for an air-handling system include ductwork. The plans for a hydronic system include more pipes and fittings.

## Unit Ventilators

At first look, the heating plans for the School Addition might appear foreign and difficult to understand, but like most of the drawings you have studied before this point, these drawings are easy to understand when taken one step at a time. Notice that each of the rooms in the addition has a rectangle on the outside wall with a callout of UV-1, UV-2, or something similar. These indicate that this building is heated by unit ventilators. **Figure 40–1** is the Schedule of Unit Ventilators. (In the actual drawing package it was #H-7 on Sheet H-4.)

## Objectives

After completing this unit, you will be able to perform the following tasks:

O Identify and briefly describe the major pieces of HVAC equipment to be used in a building.

O List the sizes of pipes and fittings shown on mechanical plans, and explain the major functions of each.

O Describe the sizes and shapes of air-handling ducts shown on mechanical drawings.

| UNIT VENTILATOR SCHEDULE | | | | | | | | | | | H-7 | | |
|---|---|---|---|---|---|---|---|---|---|---|---|---|---|
| MARK | MAKE | MODEL | FAN | | | | V/PH/HZ | HEATING | | | | | |
| | | | | | | | | HOT WATER | | | | | |
| | | | TOT CFM | MIN O.A. | RPM | HP | | EAT | EWT | GPM | MBH | ΔP | |
| UV-1 | AAF | AV-4000 | 1000 | 270 | 650 | 1/8 | 120/1/80 | 50 | 180 | 2.0 | 36 | 0.99 | |
| UV-2 | AAF | AV-5000 | 1250 | 270 | 650 | 1/8 | 120/1/60 | 50 | 180 | 2.0 | 42 | 1.27 | |
| UV-3 | AAF | AV-3000 | 750 | 250 | 650 | 1/8 | 120/1/60 | 40 | 180 | 2.0 | 28 | 0.79 | |
| UV-4 | AAF | AH-3000 | 750 | 270 | 650 | 1/8 | 120/1/60 | 40 | 180 | 2.0 | 36 | 0.79 | |
| UV-5 | AAF | AH-3000 | 750 | 270 | | 1/2 | 120/1/60 | 40 | 180 | 2.0 | 36 | 0.79 | |
| | | | | | | | | | | | | | |

| | | | |
|---|---|---|---|
| TOT CFM | total cubic feet per minute through the unit | EWT | entering water temperature |
| MIN O.A | minimum outside airflow into the unit | GPH | gallons per minute of water through the unit |
| RPM | revolutions per minute of the fan | MBH | thousands of BTUs per hour (a measure of heat output) |
| HP | fan motor horsepower | | |
| V/PH/HZ | fan motor volts, phase rating (1 or 3), and hertz (frequency or cycle) | ΔP | delta pressure (drop in water pressure through the system) |
| EAT | entering air temperature | | |

**Figure 40–1.** Unit ventilator schedule.

A *unit ventilator* is a unit that combines a means of mixing room air and outside air to ventilate the room while also warming it. The warmth is provided by the hot water that is piped in from the boiler.

**Figure 40–2** explains the principles of a unit ventilator. The heating controls include a function that controls the balance of recirculated inside air and fresh outside air. During the day, when the building is in use, the damper is opened to allow more fresh air to be circulated through the room. At night, the damper is closed, so that less cool air enters the room, resulting in an energy savings. The amount of air that is drawn in from the outside when the dampers are closed as much as possible is indicated on the drawings and on the schedule as C.F.M. (cubic feet per minute). Other information on the Schedule of Unit Ventilators describes such things as the total amount of air that passes through the unit, the water temperature intended by the designer, and the speed of the fan.

This school does not include air conditioning. If it did, the unit ventilators would be supplied with cold water when cooling is called for. The same system of balancing inside and outside air would be used for energy conservation and fresh air, but instead of passing the air over heated pipes, it would be passed over cold pipes. The Schedule of Unit Ventilators would include information for air conditioning.

The unit ventilators are only a few feet long; yet the plans show a continuous string of equipment of some sort along the exterior walls of the classrooms. Although not labeled anywhere on the plans, the equipment is bookcases; the Construction Notes refer to them. The manufacturers of unit ventilators also manufacture bookcases and filler pieces that are designed to fit with the ventilators and fill the rest of the wall space. These bookcases have a false back, with space for pipes to and from the ventilators.

In some spaces, such as stairways and storage closets, the heaters are not labeled as UV (unit ventilators). These are cabinet unit heaters (CUH) that mount on the surface of the wall and do not have ventilation capability and fan coil units (FCU) that are intended to provide more heat than ventilation. **Figure 40–3** shows the schedules for the units marked as CUH, UH, and FCU.

## Heating Piping

Once the basic type of heating system has been determined, the next step is to understand the piping that carries water to and from the heating units. It is easy to spot the pipes on the heating plan. A solid line indicates supply pipes, and a dashed line indicates return pipes. Pipe sizes are indicated by callouts throughout the plans. Fittings such as valves, elbows, and tees are represented by symbols. **Figure 40–4** shows the most common symbols for mechanical systems. There are standard piping symbols,

EXTERIOR WALL
(THICKNESS
YARDS)

BLENDED
HEATED
AIR

FINS TO
RADIATE HEAT

RECIRCULATED
INSIDE AIR

OUTSIDE
AIR

FLOOR SYSTEM

PIPES CARRY
HOT WATER

CONTROL DAMPER OPENS DURING USE
HOURS TO PROVIDE FRESH AIR AND
CLOSES AT NIGHT TO CONSERVE ENERGY

**Figure 40–2.** Operation of a unit ventilator.

| MARK | MAKE | MODEL | STYLE & ARRANGEMENT | STEAM | | WATER | | | | CFM | RPM | EAT | MBH | ELECTRICAL | |
| | | | | PSIG | IB/HR | EWT | △T | GPM | △P" | | | | | V/ø/HZ | HP |
|---|---|---|---|---|---|---|---|---|---|---|---|---|---|---|---|
| CUH–1 | STERLING | RWI–1130–04 | RECESSED WALL, INVERTER | | | 180 | 34 | 2.0 | 0.44 | 420 | 1050 | 60 | 33.8 | 120/1/60 | 1/10 |

## CABINET UNIT HEATER SCHEDULE

| MARK | MAKE | MODEL | TYPE | WATER | | | MBH | CFM | EAT | LAT | MOTOR | | |
| | | | | ENT. | GFM | △P | | | | | RPM | H.P. | V/ø/HZ |
|---|---|---|---|---|---|---|---|---|---|---|---|---|---|
| UH–1 | MODINE | HS–18L | HORIZONTAL | 180 | 1.1 | 0.4 | 9.4 | 364 | 60 | 84 | 1550 | 16MHP | 120/1/60 |

## UNIT HEATER SCHEDULE

| MARK | MAKE | MODEL | CFM & TOT./F.A. | EXT. S.P. | ELECTRICAL | | HEATING | | | | |
| | | | | | WATTS | VOLTS/PH/HZ | EAT | EWT | GPM | MBH | △P |
|---|---|---|---|---|---|---|---|---|---|---|---|
| FCU–1 | AAF | SFG–TFA–3000 | 350 | 0 | 110 | 120/1/60 | 50 | 180 | 1.0 | 18 | 0.1 |

## FAN COIL UNIT SCHEDULE

**Figure 40–3.** Schedules of cabinet unit heaters, unit heaters, and fan cabinet unit heaters.

## MECHANICAL SYMBOLS

| DESCRIPTION | SYMBOL |
|---|---|
| RECTANGULAR OR SQUARE SUPPLY DUCT TURNED UP | |
| RECTANGULAR OR SQUARE SUPPLY DUCT TURNED DOWN | |
| RECTANGULAR OR SQUARE EXHAUST DUCT TURNED UP | |
| RECTANGULAR OR SQUARE EXHAUST DUCT TURNED DOWN | |
| ROUND SUPPLY DUCT TURNED UP | |
| ROUND SUPPLY DUCT TURNED DOWN | |
| RECTANGULAR CEILING DIFFUSER | |
| ROUND CEILING DIFFUSER | |
| FLEXIBLE CONNECTION | |
| TRANSITION: FOT = FLAT ON TOP | FOT |

## MECHANICAL SYMBOLS (CONTINUED)

| DESCRIPTION | SYMBOL |
|---|---|
| ELECTRIC OPERATED DAMPER | E.O.D. |
| FIRE DAMPER | F.D. |
| SMOKE DAMPER | S.D. |
| ELECTRIC HEATER IN DUCT | |
| SUPPLY OUTLET WITH SIZE AND AIR QUANTITY SHOWN | 12 X 8 200 CFM |
| DEFLECTOR IN DUCT BEHIND REGISTER OR GRILLE (ARROW INDICAES DIRECTION OF FLOW) | |
| TURNING VANES IN A SQUARE THROAT ELBOW | |
| TURNING VANES IN A ROUND THROAT ELBOW | |
| PLAN VIEW OF TRANSITION | |
| OFF-SET UP IN DIRECTION OF ARROW | |
| DUCT DIMENSIONS—FIRST FIGURE IS THE SIDE OF DUCT SHOWN (12 X 10) | |

**Figure 40–4.** Mechanical and plumbing symbols.

## MECHANICAL SYMBOLS (CONTINUED)

| DESCRIPTION | SYMBOL |
|---|---|
| ACCOUSTICAL LINING INSIDE INSULATION | |
| BRANCH TAP IN DUCT | |
| SPLITTER FITTING WITH DAMPER | S.D. |
| VOLUME-DAMPER | V.D. |
| BACKDRAFT DAMPER | BDD |
| ACCESS DOOR IN DUCT 10"X 10" SIZE | 10 x 10  A.D. |
| PNEUMATIC OPERATED DAMPER | P.O.D. |
| THREE-WAY VALVE | |
| PRESSURE REDUCING VALVE | |
| PRESSURE RELIEF VALVE OR SAFETY VALVE | |
| SOLENOID VALVE | |
| PIPE TURNED UP (ELBOW) | |
| PIPE TURNED DOWN (ELBOW) | |
| TEE (OUTLET UP) | |
| TEE (OUTLET DOWN) | |

## MECHANICAL SYMBOLS (CONTINUED)

| DESCRIPTION | SYMBOL |
|---|---|
| BACKFLOW PREVENTER | B.F.P. |
| UNION | |
| REDUCER | |
| CHECK VALVE | FLOW |
| GATE VALVE | OR |
| GLOBE VALVE | OR |
| BALL VALVE | |
| BUTTERFLY VALVE | |
| DIAPHRAGM VALVE | |
| ANGLE GATE VALVE | |
| ANGLE GLOBE VALVE | |
| PLUG VALVE | |
| LOW PRESSURE STEAM | LPS |
| LOW PRESSURE CONDENSATE | LPC |
| PUMPED CONDENSATE | PC |
| FUEL OIL SUPPLY | FOS |
| FUEL OIL RETURN | FOR |
| HOT WATER SUPPLY | HWS |
| HOT WATER RETURN | HWR |
| COMPRESSED AIR | A |
| REFRIGERANT SUCTION | RS |

**Figure 40–4.** (continued)

## MECHANICAL SYMBOLS (CONTINUED)

| DESCRIPTION | SYMBOL |
|---|---|
| REFRIGERANT LIQUID | — RL — |
| REFRIGERANT HOT GAS | — RHG — |
| CONDENSATE DRAIN | — CD — |
| FUEL GAS | — G — |
| CHILLED WATER SUPPLY | — CWS — |
| CHILLED WATER RETURN | — CWR — |

## PLUMBING SYMBOLS

| DESCRIPTION | SYMBOL |
|---|---|
| METER | M |
| SPRINKLER PIPING | S |
| SPRINKLER HEAD | |
| FLOOR DRAIN | F.D. |
| CLEAN-OUT | C.O. |
| TUB | |
| TANK-TYPE WATER CLOSET | |
| WALL-MOUNTED LAVATORY | |
| URINAL | |
| SHOWER | |
| WATER HEATER | WH |
| MANHOLE | MH |

## PLUMBING SYMBOLS (CONTINUED)

| DESCRIPTION | SYMBOL |
|---|---|
| WALL HYDRANT | |
| YARD HYDRANT | Y.H. |
| FLUSH VALUE WATER CLOSET | |
| COUNTER-TYPE LAVATORY | |
| KITCHEN SINK (DOUBLE BOWL) | |
| SOIL, WASTE OR DRAIN LINE | |
| PLUMBING VENT LINE | – – – – |
| COLD WATER (DOMESTIC) | — – – |
| HOT WATER (DOMESTIC) | — – – – |
| HOT WATER RETURN (DOMESTIC) | — – – – |
| FIRE LINE | — F — |
| FUEL GAS LINE | — G — |
| ACID WASTE LINE | — AW — |
| VACUUM LINE | — V — |
| COMPRESSED AIR LINE | — A — |
| BACKFLOW PREVENTER | BFP |
| GATE VALVE | OR |
| GLOBE VALVE | OR |
| CHECK VALVE(ARROW INDICATES DIRECTION OF FLOW) | |

**Figure 40–4.** (continued)

## PLUMBING PIPE SYMBOLS (CONTINUED)

| DESCRIPTION | SYMBOL |
|---|---|
| UNION | |
| PIPE TURNED DOWN | |
| PIPE TURNED UP | |
| TEE OUTLET UP | |
| TEE OUTLET DOWN | |
| TEE OUTLET TO SIDE | |
| REDUCER | |
| PIPE SLEEVE | |

**Figure 40–4.** (continued)

**Figure 40–5.** This isometric drawing shows how the heating pipes drop down from the ceiling to the level of the ventilators.

but each drafter uses a few special representations of his or her own, so a symbol legend is included with the drawings.

Pipes are easier than ducts to coordinate with structural elements because the direction of the pipe can be changed easily and frequently. Pipes might run horizontally in a ceiling for some distance, then loop around a column and drop down to a unit ventilator. It is not practical to show the layout of all pipes with separate plan and riser views. The common method for showing pipes where both horizontal and vertical layout is involved is to use isometric schematics. Plumbing isometrics are discussed in Unit 35. Sheet H-1 of the School Addition drawings combines plan views and isometric schematics. The long runs of pipes are shown in plan view to indicate where the pipes are in relation to the walls. Where the pipes need to drop down from the ceiling to the level of the ventilators, they are shown in isometric views, **Figure 40–5.**

For the sake of explanation, the following discussion traces the hot-water supply piping (broken line) in room 109, the storage room on the first floor, E-7 on Sheet H-1. Most of what is discussed here is also shown in **Figure 40–5,** but we will start closer to the center of the building, where the branch tees to the right (north) off the main supply running from top to bottom of the drawing. See the symbol legend A-11 on Sheet H-1 to interpret the symbols as you trace the piping.

The first thing we encounter is a valve, then an elbow toward the top of the drawing, and then another elbow back to the right. Construction Note 13 indicates that the pipes pass through a structural steel member. The line is broken for clarity through the area where ductwork is shown that would make reading the piping drawing difficult. Pick the line up again in the storage room 109. Next we come to a tee where the pipe drops down to the level of the unit ventilator. If we continued on to the right or north, we would come to Construction Note 2, which tells us that the symbol indicates a riser to the second floor. The riser is not shown in the wall, where it would actually be positioned, because doing so would bury the riser symbol in the wall with material symbols and other lines. The pipe going down is actually shown at an angle up and to the right. This is the isometric portion of the drawing. At the level of the unit ventilators, the pipes are arranged in a special way to eliminate noises that might result from expansion and contraction as the water temperature changes, **Figure 40–6.** The ends of the pipes are shown with break lines, which indicate that the pipes actually continue to their obvious destinations: fittings on the unit ventilators.

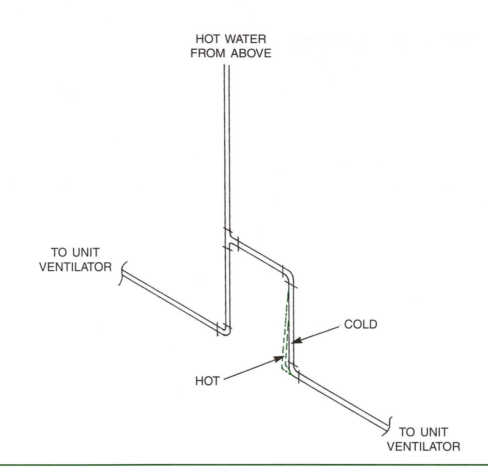

HOT WATER
FROM ABOVE

TO UNIT
VENTILATOR

COLD

HOT

TO UNIT
VENTILATOR

**Figure 40–6.** Noises are eliminated by allowing pipes to expand and contract.

# Air-Handling Equipment

The unit ventilators introduce fresh air into the school addition. A separate ventilation system removes stale air from each room of the school. All the rooms in the addition except the smallest closets and the stairways are provided with a system of ducts and a fan to ensure air circulation. A *louvered grille* in the room receives the air and channels it into a duct that carries it to the roof, where a fan exhausts it to the outside.

**Figure 40–7** explains the designations for outlet and inlet grilles. The ducts that carry the air to the roof vent are shown by their outlines and do not indicate their sizes. A square or rectangle with a diagonal line through it indicates where the duct rises to the next floor or the roof. Several notes indicate that ducts rise up to EF-# on the roof. *EF* is an abbreviation for exhaust fan. Some of the exhaust fans occupy several square feet of space on the roof, so they cannot be placed too close together.

RB – 6x6 – 270

CFM
SIZE
S = SUPPLY/R = RETURN

**Figure 40–7.** Explanation of grille designations.

Trace the flow of air as it is exhausted from the janitor's closet on the first floor. The path starts with a louver marked RB (return) – 8 × 6 (cross-sectional size in inches) – 75 (cubic-feet-per-minute nominal airflow). Between the janitor's closet and the storage room, the air enters a 6 × 6 duct, which goes up through a chase (space between two walls) to the second floor. At the second floor it is joined by the exhaust air from the second floor janitor's closet. On the Second Floor Plan there is also a symbol indicating a fire damper designated as W/ Access Door At Floor. This damper

TOP DAMPER BLADE IS
RIVETED TO TOP OF FRAME

FRAME MOUNTS IN
THRU-WALL DUCT

FUSIBLE LINK HOLDS
DAMPER BLADES IN THE
UP (OPEN) POSITION.

FUSIBLE LINK HAS
MELTED AND ALLOWED
DAMPER BLADES TO FALL
TO CLOSED POSITION

NORMAL POSITION

FIRE (CLOSED) POSITION

**Figure 40–8.** Fire damper for ductwork.

(metal gate) closes off the duct if it gets hot in a fire, **Figure 40–8.** This is necessary because the building code requires fire-rated construction, and an open duct is a great place for fire to spread. The access door at the floor allows for maintenance and repairs to the damper. Notes at both levels indicate that the duct between floors is 6 × 6. Above the second floor louver, the duct is 8 × 8; and at the second floor ceiling, it is offset an unspecified distance into the storage room, where it turns up to EF-4 on the roof. The reason for the offset is because it would otherwise be too close to the other duct such that there would not be enough room for both exhaust fans on the roof.

## Plumbing

The plumbing for the School Addition consists of DWV (drainage, waste, and vent) piping to remove effluent (wastewater), hot- and cold-water distribution, and fixtures. The DWV system is more complex for a building

this size than for a house, because the large flat roof collects quite a bit of water in a rainstorm and there must be a way to get rid of it. Therefore, the Second Floor Plumbing Plan shows roof drains, marked R.D., and storm drain piping, indicated by a heavy line. Callouts along the storm drain indicate the size of the pipe and its slope and give information about how the storm drain is to be coordinated with structural components of the building. On the First Floor Plumbing Plan the storm drain is represented by a heavy dashed line, indicating that it is below grade.

Notice that the *storm drain* and the *sanitary drain* are separate systems. The storm drain only needs to carry rainwater away from the building. The sanitary drain must connect to a sewer or septic system. Although the storm drain and the sanitary drain pipes cross near the janitor's closet on the first floor, they are not connected.

In the region where this building was constructed, drainage pipes outside the building lines are installed by the site contractor, not the plumbing contractor. Therefore, the plumbing drawings only show the storm drain and sanitary drain to a point 5'-0" outside the building lines. A note where each line passes through the building wall indicates the *invert elevation* (elevation of the bottom of the inside of the pipe).

The hot- and cold-water piping is similar to that found in homes, except that the hot water is a two-pipe system. A return hot-water pipe allows the hot water to circulate continuously, ensuring that hot water is readily available at the point of use. The hot-water supply and hot-water return can be distinguished from each other by the number of short dashes in the symbol. These are explained on the symbol legend in the drawing packet. The hot-water return, which is usually smaller than the supply, maintains circulation to a point close to where most of the hot water will be used. It is often not practical to continue a hot-water return pipe to every fixture in an area with several fixtures like the School Addition toilet rooms. **Figure 40–9** shows riser diagrams for the fixture connections in the School Addition. Notice that there is no hot-water return shown on these diagrams, because the return is connected to the supply before the supply reaches the first fixture.

A plumbing riser diagram is only used to indicate design intent and does not indicate a piping route. The sizes indicated by the engineer are minimum sizes. Company preferences and local codes can dictate that larger sizes be used. Notice that Riser 1 and Riser 4 in **Figure 40–9** are identical. This does not mean that they will always be installed the same way. The notes on Riser 1 are different from the notes on Riser 4. The notes on both risers are relevant for Risers 1 and 4, and the vacuum breakers required on S-1 in Riser 1 are also required on S-1 fixtures in Riser 4. The note to route all piping in the casework in Riser 4 is also relevant for the piping in Riser 1.

It is important for plumbers to know and comply with local codes. In Riser 3, the vent serving MB-1 is offset 90° below the fixture. That is an illegal installation by most codes, and the plumber would install the **vent pipe** in accordance with local codes and not use the riser diagram as a guide for routing the pipe. Where the pipes seem to connect but are actually different piping systems, you can outline the pipe route with different color pencils to clarify the different piping systems.

Figure 40–9. Riser diagrams for school fixtures.

## ✓ CHECK YOUR PROGRESS

Can you perform these tasks?

● Locate information on schedules of equipment, and relate that information to the plans.

● Understand the symbols and abbreviations used for air-conditioning, heating, and plumbing drawings.

● Explain the sizes and shapes of ducts to be used for air handling and ventilation.

● Trace the path of piping, and identify the sizes of pipes and fittings used.

## ASSIGNMENT

Refer to the drawings of the School Addition (included in the packet) to complete this assignment.

1. How many unit ventilators are shown on the first floor?

2. How is the machinery room on the first floor behind the elevator heated?

3. List in order the sizes of pipe that the water flows through, starting from the point where the water enters the addition to get to the heating unit in classroom 210 and then back to the point where it leaves the addition.

4. What are the dimensions of the grille where stale air is vented out of classroom 209?

5. List in order the sizes of the ductwork that exhaust air passes through from the point where the air leaves classroom 209 until it is outside the building.

6. List in order the sizes of the ductwork that exhaust air passes through from the point where the air leaves the girls' toilet room on the first floor until it exits the building.

7. How many gallons per minute of water are expected to flow through the heating unit in stair #2?

8. Why are two roof drains shown on the First Floor Plumbing Plan of this two-story building?

9. What is designated as EWC-1 on the plumbing plans?

10. What is the vertical distance between the storm drain and the sanitary drain at the point where they cross?

11. What is the diameter of the pipe that is used for the storm drain where it goes from the second floor to the first floor?

12. What size pipe is used for the domestic hot-water return line?

13. Where does the domestic hot-water return line tee out of the domestic hot-water supply line?

14. Where would you shut off the cold-water supply to the kitchenette unit in room 101 without shutting off the cold water to the toilet rooms on the same floor?

# Electrical Drawings

## Electrical Drawings

Electrical drawings deal predominantly with circuits. Knowledge and under-standing of the basic circuits used in buildings is necessary for those working with and interpreting electrical drawings. The four basic methods of showing electrical circuits are:

1. Plan views

2. Single-line diagrams

3. Riser diagrams

4. Schematic diagrams

## Plan Views

The electrical floor plan of a building shows all the exterior walls, interior parti-tions, windows, doors, stairs, cabinets, and so on along with the location of the electrical items and their circuitry.

## Power Circuits

The *power circuit* electrical floor plan shows electrical outlets and devices and includes duplex outlets, specialty outlets, telephone, fire alarm, etc. The conventional method of showing power circuits is to use long dash lines when specified to be installed in the slab or underground and to use solid lines when concealed in ceilings and walls. Short dash lines indicate exposed wir-ing. Solid lines are often used in commercial installations when the raceway (conduit) is to be exposed (surface mounted). Slash marks through the circuit lines are used to indicate the number of conductors. Full slash marks are the circuit conductors, a longer full slash is the neutral, and half slash marks are the ground wires. When only two wires are required, no slash marks are used. The typical wire size for commercial construction is No. 12 AWG. The arrows indicate "home runs" to the designated panel. The panel and circuit number designations are adjacent to the arrowheads. **Figure 41–1** lists the typical

## Objectives

After completing this unit, you will be able to perform the following tasks:

O Explain the information found on a lighting plan.

O List the equipment served by an individual branch circuit using electrical plans, riser drawings, and schedules.

O Explain the information on a schematic diagram.

## CIRCUITING

——1→→2  BRANCH-CIRCUIT HOME RUN TO PANEL*

——/// —  THREE WIRES IN CABLE OR RACEWAY

——//// —  FOUR WIRES IN CABLE OR RACEWAY, ETC.

——/// —  SOME DRAWINGS SHOW THIS METHOD OF CONDUCTOR IDENTIFICATION:

    EQUIPMENT GROUNDING CONDUCTOR: LONG LINE WITH DOT.

    NEUTRAL CONDUCTOR: LONG LINE.

    PHASE CONDUCTOR WITH SWITCH LEGS: SHORT LINE.

————————  WIRING CONCEALED IN CEILING OR WALL

– – – – – –  WIRING CONCEALED IN FLOOR

– — – — –  WIRING EXPOSED

————○  WIRING TURNED UP

————●  WIRING TURNED DOWN

————CO  CONDUIT ONLY (EMPTY)

⌒  SWITCH LEG INDICATION. CONNECTS OUTLETS WITH CONTROL POINTS.

* AN ARROW INDICATES A BRANCH-CIRCUIT HOME RUN TO PANEL.

THE NUMBER OF ARROWS INDICATES THE NUMBER OF CIRCUITS.

IF THERE ARE NO CROSSHATCHES, THEN IT IS ASSUMED THAT THE RACEWAY CONTAINS TWO WIRES.

**Figure 41–1.** Circuiting symbols.

circuiting symbols. The electrical installer will normally group home runs in a raceway with combinations similar to the following:

O Three-phase systems with three circuit wires and a common neutral in a raceway

O Single-phase systems with two circuit wires and a common neutral in a raceway

Note:
To reduce the harmonic problems caused by solid-state devices (nonlinear loads), many electrical designers are requiring separate neutrals with each phase conductor.

Some equipment, such as television, fire alarm, clock, PA, and sound, might not be connected with wiring lines on the plan. This indicates that wiring for that equipment is not part of the general circuit wiring and is probably not to be fed through the same panels as the rest of the electrical equipment. For example, on the School Addition the PA system and the clocks are fed from the sound system console and do not involve the general circuit wiring.

## Lighting Circuits

There is usually a *lighting circuit* electrical floor plan for each level or major space within a building. This floor plan shows light fixtures, emergency lighting, security lighting, and special lighting control (i.e., photocell, motion detector, etc.). Typically included is a reflected ceiling plan, **Figure 41–2.** That means that it is a plan view but shows what is on the ceiling as though it were reflected onto the floor plan, **Figure 41–2.** The reflected ceiling plan shows each light fixture with a circle, square, or rectangle that approximates the shape of the fixture. The lighting circuitry on the School Addition is shown with three types of wiring circuit lines, solid for unswitched, dotted for switched, and line-dash-line for the motion sensor circuits. The conventional method of showing lighting circuits is the same as previously explained in the paragraph above on power circuits.

*Motion sensors* are an interesting feature of the lighting design for the School Addition. The motion sensors are shown as small circles containing the letter *M.* A motion sensor detects any movement in the area of the sensor, so that in a room controlled by a motion sensor, if there is no movement in the room for a period of time, the sensor opens its contacts, like turning a switch off. A motion detector can be used together with a switch so that the lights are controlled normally by the switch. But if someone forgets to turn the lights off, the motion detector will.

A low-voltage lighting control system requires a *relay* that is activated by the light switch. The relay opens or closes the higher-voltage fixture, **Figure 41–3.** The relays on the School Addition are indicated by the letter *R* in a small square. Notice that every lighting switch is connected to a relay.

CEILING OF TWO ROOMS

REFLECTED CEILING PLAN

**Figure 41–2.** Reflected ceiling plan.

## Symbols

Electrical *symbols* are used to simplify the drafting and later the interpreting of the drawings. Electrical symbols *are not standardized* throughout the industry. Most drawings will have a symbol legend or list. You must be knowledgeable of the symbols specifically used on each project since designers modify basic symbols to suit their own needs. Many symbols are similar (circle, square, etc.). The addition of a line, dot, shading, letters, numbers, and so forth gives the specific meaning to the symbol. Learning the basic form of the various symbols is the best starting point in developing the ability to interpret the drawings and their related symbol meanings. **Figure 41–4** lists common electrical symbols.

The School Addition drawing E-1, symbol legend A-4, contains an electrical symbol list for the project. From the symbol list, it can be seen that there are duplex receptacles, switches, telephone outlets, special-purpose outlets, and fire alarm devices mounted at various heights. Then there is the General Note in the symbol legend that specifies that all mounting heights are to be verified and "modified as directed." This is an example of why the installer must become familiar with the drawings and specifications far in advance of the scheduled time for the installation. The installer must request clarification or direction and give the designers reasonable time to clarify the questionable specified instructions.

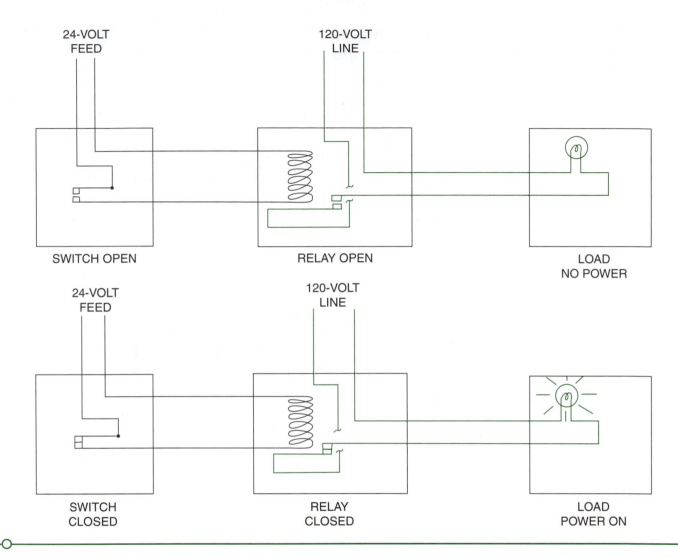

**Figure 41–3.** A relay allows a low-voltage circuit to start and stop a higher- voltage circuit.

## Single-Line Diagrams

The electrical service *single-line diagram* on the School Addition is shown on drawing E-2, Single-Line Diagram E-17. The electric power is brought into the school by way of the service entrance section (S.E.S.) The note "—coordinate with the local utility requirements prior to rough-in" indicates that the electrical contractor is to furnish and install the three 4-inch conduits and that the utility company wants to inspect the conduit installation prior to backfilling the trench where the conduits are installed; however, this should be verified at a preconstruction meeting with the utility company. There are times when an electrical service, or electrical distribution raceway,

would require being concrete encased for protection. This type of concrete-encased underground raceway system is called a *duct bank,* **Figure 41–5.**

The note does not specifically indicate who furnishes and installs the utility feeders in the three 4-inch conduits. This could be an expensive item and should be clarified with the utility company prior to estimating and bidding the project.

To comply with the *National Electrical Code® (NEC®),* the service entrance section may have up to, but must not exceed, six main disconnecting units (switches or circuit breakers). Four main fused switches are shown in the EXISTING portion of the single-line drawing. Note that this single-line drawing shows only a portion of the

## Electrical Reference Symbols

*ELECTRICAL ABBREVIATIONS*
(Apply only when adjacent to an electrical symbol.)

| | |
|---|---|
| Central Switch Panel | CSP |
| Dimmer Control Panel | DCP |
| Dust Tight | DT |
| Emergency Switch Panel | ESP |
| Empty | MT |
| Explosion Proof | EP |
| Grounded | G |
| Night Light | NL |
| Pull Chain | PC |
| Rain Tight | RT |
| Recessed | R |
| Transfer | XFER |
| Transformer | XFRMR |
| Vapor Tight | VT |
| Water Tight | WT |
| Weather Proof | WP |

*ELECTRICAL SYMBOLS*

Switch Outlets

| | |
|---|---|
| Single-Pole Switch | S |
| Double-Pole Switch | $S_2$ |
| Three-Way Switch | $S_3$ |
| Four-Way Switch | $S_4$ |
| Key-Operated Switch | $S_K$ |
| Switch and Fusestat Holder | $S_{FH}$ |
| Switch and Pilot Lamp | $S_P$ |
| Fan Switch | $S_F$ |
| Switch for Low-Voltage Switching System | $S_L$ |
| Master Switch for Low-Voltage Switching System | $S_{LM}$ |
| Switch and Single Receptacle | S |
| Switch and Duplex Receptacle | S |
| Door Switch | $S_D$ |
| Time Switch | $S_T$ |
| Momentary Contact Switch | $S_{MC}$ |
| Ceiling Pull Switch | S |
| "Hand-Off-Auto" Control Switch | HOA |
| Multi-Speed Control Switch | M |
| Push Button | • |

**Receptacle Outlets**

Where weather proof, explosion proof, or other specific types of devices are to be required, use the upper-case subscript letters. For example, weather proof single or duplex receptacles would have the uppercase WP subscript letters noted alongside of the symbol. All outlets should be grounded.

Single Receptacle Outlet

Duplex Receptacle Outlet

Triplex Receptacle Outlet

Quadruplex Receptacle Outlet

**Figure 41–4.** Recommended electrical symbols.

| | |
|---|---|
| Duplex Receptacle Outlet (Split Wired) | |
| Triplex Receptacle Outlet (Split Wired) | |
| 250-V Receptacle Single Phase. Use Subscript Letter to Indicate Function (DW—Dishwasher, RA—Range, CD—Clothes Dryer) or Numeral (with explanation in symbol schedule). | |
| 250-V Receptacle Three Phase | |
| Clock Receptacle | C |
| Fan Receptacle | F |
| Floor Single Receptacle Outlet | |
| Floor Duplex Receptacle Outlet | |
| Floor Special-Purpose Outlet | * |
| Floor Telephone Outlet (Public) | |
| Floor Telephone Outlet (Private) | |

Example of the use of several floor outlet symbols to identify a 2, 3, or more gang flow outlet.

Underfloor duct and junction box for triple, double, or single duct system as indicated by the number of parallel lines.

Example of use of various symbols to identify location of different types of outlets or connections for underfloor duct or cellular floor systems.

Cellular Floor Header Duct

*Use numeral keyed to explanation in drawing list of symbols to indicate usage.

## Circuiting

| | |
|---|---|
| Wiring Exposed (Not in Conduit) | —— E —— |
| Wiring Concealed in Ceiling or Wall | |
| Wiring Concealed in Floor | |
| Wiring Existing* | |
| Wiring Turned Up | |
| Wiring Turned Down | |
| Branch Circuit Home Run to Panel Board | 2  1 |

Number of arrows indicates number of circuits. (A number of each arrow may be used to identify circuit number.)‡

### Bus Ducts and Wireways

| | |
|---|---|
| Trolley Duct‡ | T  T |
| Busway (Service, Feeder, or Plug-in)‡ | B  B |
| Cable Trough Ladder or Channels‡ | C  C |
| Wireway‡ | W  W |

### Panelboards, Switchboards, and Related Equipment

| | |
|---|---|
| Flush-Mounted Panelboard and Cabinet‡ | |
| Surface-Mounted Panelboard and Cabinet‡ | |
| Switchboard, Power Control Center, Unit Substations (Should be drawn to scale.)‡ | |
| Flush-Mounted Terminal Cabinet (In small scale drawings the TC may be indicated alongside the symbol.)‡ | TC |
| Surface-Mounted Terminal Cabinet (In small scale drawings the TC may be indicated alongside the symbol.)‡ | TC |

**Figure 41–4.** (continued)

| | |
|---|---|
| Pull Box (Identify in Relation to Wiring System Section and Size.) | ▦ |
| Motor or Other Power Controller (May Be a Starter or Contactor.)‡ | ⊠ |
| Externally-Operated Disconnection Switch | |
| Combination Controller and Disconnection Means | ⊠ |

## Power Equipment

| | |
|---|---|
| Electric Motor (hp As Indicated) | 1/4 |
| Power Transformer | |
| Pothead (Cable Termination) | |
| Circuit Element (e.g., Circuit Breaker) | CB |
| Circuit Breaker | |
| Fusible Element | |
| Single-Throw Knife Switch | |
| Double-Throw Knife Switch | |
| Ground | ⊥⏐⏐ |
| Battery | |
| Contactor | C |
| Photoelectric Cell | PE |
| Voltage Cycles, Phase | Ex: 480/60/3 |
| Relay | R |
| Equipment Connection (As Noted) | ▲ |

*Note: Use heavy weight line to identify service and leaders.
Indicate empty conduit by notation CO (conduit only).
‡ Note: any circuit without further identification indicates two-wire circuit for a greater number of wires, indicate with cross lines, e.g.:

╫ 3 wires        ╫╫ 4 wires

---

Neutral wire may be shown longer. Unless indicated otherwise, the wire size of the circuit is the minimum size required by the specification. Identify different functions of wiring system, e.g. signaling system by notation or other means.
‡ Identify by notation or schedule.

## Remote Control Stations for Motors or Other Equipment

| | |
|---|---|
| Push Button Station | PB |
| Float Switch (Mechanical) | F |
| Limit Switch (Mechanical) | L |
| Pneumatic Switch (Mechanical) | P |
| Electric Eye (Beam Source) | ↘↘ |
| Electric Eye (Relay) | ↘↘ |
| Temperature Control Relay Connection (3 Denotes Quantity.) | R ₃ |
| Solenoid Control Valve Connection | S |
| Pressure Switch Connection | P |
| Aquastat Connection | A |
| Vacuum Switch Connection | V |
| Gas Solenoid Valve Connection | G |
| Flow Switch Connection | F |
| Timer Connection | T |
| Limit Switch Connection | L |

## Lighting

| | Ceiling | Wall |
|---|---|---|
| Surface or Pendant Incandescent Fixture (PC = Pull Chain) | TYPE ◯ WATTS | SWITCH ─◯ PC CIRCUIT |
| Surface or Pendant Exit Light | ⊗ | ─⊗ |
| Blanked Outlet | Ⓑ | ─Ⓑ |
| Junction Box | Ⓙ | ─Ⓙ |

**Figure 41–4.** (continued)

Electrical Drawings 317

Recessed Incandescent Fixtures

Surface or Pendant Individual
Fluorescent Fixture

Surface or Pendant Continuous-
Row Fluorescent Fixture
(Letter Indicating Controlling Switch)

Bare-Lamp Fluorescent Strip*

*In the case of continuous-row bare-lamp fluorescent strip
above an area-wide diffusing means, show each fixture run
using the standard symbol; indicate area of diffusing means
and type by light shading and/or by light shading and/or
drawing notation.

## Electric Distribution or Lighting System, Aerial

Pole‡

Steel or Parking Lot Light
and Bracket‡

Transformer‡

Primary Circuit‡

Secondary Circuit‡

Down Guy

Head Guy

Sidewalk Guy

Service Weather Head‡

## Electric Distribution or Lighting System, Underground

Manhole‡

Handhole‡

Transformer Manhole
or Vault‡

Transformer Pad‡

Underground Direct, Burial Cable
(Indicate type, size, and number
of conductors by notation
or schedule.)

Underground Duct Line
(Indicate type, size, and
number of ducts by cross-
section identification of each
run by notation or schedule.
Indicate type, size, and number
of conductors by notation or
schedule.)

Street Light Standard Feed From
Underground Circuit‡

‡ Identify by notation or schedule.

Signaling System Outlets

## Institutional, Commercial, and Industrial Occupancies

I. Nurse Call System Devices
(Any Type)

Basic Symbol

(Examples of individual item identification. Not a part of standard.)

Nurses' Annunciator
(Adding a number after it
indicates number of lamps,
e.g., +①24.)

Call Station, Single Cord,
Pilot Light

Call Station, Double Cord,
Microphone Speaker

Corridor Dome Light, 1 Lamp

Transformer

Any Other Item on Same System
(Use Numbers as Required.)

II. Paging System Devices
(Any Type)

Basic Symbol

**Figure 41–4.** (continued)

(Examples of individual item identification. Not a part of standard.)

Keyboard      ⊣◇ 1

Flush Annunciator      ⊣◇ 2

Two-Face Annunciator      ⊣◇ 3

Any Other Item on Same System
(Use Numbers as Required.)      ⊣◇ 4

III. Fire Alarm System Devices
(Any Type) Including Smoke and
Sprinkler Alarm Devices

Basic Symbol

(Examples of individual item identification. Not a part of standard.)

Control Panel      ⊣□ 1

Station      ⊣□ 2

10" Gong      ⊣□ 3

Presignal Chime      ⊣□ 4

Any Other Item on Same System
(Use Numbers as Required.)      ⊣□ 5

IV. Staff Register System Devices
(Any Type)

Basic Symbol

(Examples of individual item identification. Not a part of standard.)

Phone Operators' Register      ⊣◇ 1

Entrance Register (Flush)      ⊣◇ 2

Staff Room Register      ⊣◇ 3

Transformer      ⊣◇ 4

Any Other Item on Same System
(Use Number as Required.)      ⊣◇ 5

V. Electric Clock System Devices
(Any Type)

Basic Symbol

(Examples of individual item identification. Not a part of standard.)

Master Clock      ⊣⬡ 1

12" Secondary (Flush)      ⊣⬡ 2

12" Double Dial (Wall Mounted)      ⊣⬡ 3

18" Skeleton Dial      ⊣⬡ 4

Any Other Item on Same System
(Use Numbers as Required.)      ⊣⬡ 5

VI. Public Telephone System Devices

Basic Symbol

(Examples of individual item identification. Not a part of standard.)

Switchboard      ◀ 1

Desk Phone      ◀ 2

Any Other Item on Same System
(Use Numbers as Required.)      ◀ 3

VII. Private Telephone System Devices
(Any Type)

Basic Symbol

(Examples of individual item identification. Not a part of standard.)

Switchboard      ◁ 1

Wall Phone      ◁ 2

Any Other Item on Same System
(Use Numbers as Required.)      ◁ 3

VIII. System Devices
(Any Type)

Basic Symbol

(Examples of individual item identification. Not a part of standard.)

**Figure 41–4.** (continued)

**Figure 41–5.** Duct bank system.

complete existing service entrance section. The service entrance section indicates that it is located outside by the NEMA 3R ENCLOSURE definition. NEMA, National Electric Manufacturers Association, generates specifications that are recognized as design standards for electrical boxes, devices, and equipment. NEMA 3R is a rain-tight designation. The NEW portion of the single-line drawing shows the electrical distribution to be installed in this School Addition. The feeders for the new elevator, electrical panels L10 and L11, are fed from existing panel MDP. These new electrical loads are:

1. 60-amp circuit breaker with four #4 AWG, one #10 ground, and one #10 isolated ground in a 1¼-inch conduit feeding the elevator

2. 100-amp circuit breaker with four #1 AWG and one #6 ground in a 1½-inch conduit feeding panel L11

3. 100-amp circuit breaker with four #1 AWG and one #6 ground in a 1½-inch conduit feeding panel L10

A complete building electrical floor plan is not a part of the School Addition drawings, and the EXISTING electrical distribution plan on drawing E-2 shows only a portion of the total electrical distribution section. This required notes 6, 7, and 8 to be added to drawing E-2, Single-Line Diagram E-17, giving the contractor the lengths of the feeders for the elevator, panels L10 and L11, which are to be included in this School Addition. In a typical design, a complete building electrical plan is

provided indicating the locations of all existing and new electrical equipment. The feeder length would be determined from that drawing.

## Riser Diagrams

A *riser diagram* is so named because it usually shows the path of wiring or raceway from one level of a building to another and because the wiring rises from one floor to the next. A riser diagram does not give information about where equipment is to be located in a room or area. Riser diagrams are used because they are particularly easy to understand. Therefore, they do not require much explanation.

A *power riser diagram,* **Figure 41–6,** shows a typical building's electrical service and related components. This figure is not the same electrical service as the School Addition Single-Line Diagram, E-17, on drawing E-2, but comparing the two diagrams shows how a power riser diagram greatly simplifies the interpretation of an installation drawing.

A *special riser diagram* is used for many systems that include:

1. Fire alarm

2. Security

3. Telephone

4. Clock

**Figure 41–6.** Typical riser diagram showing an overview of the building's electrical service and related components.

5. Signal
   - Bell
   - Call (nurse, emergency, etc.)
   - Water sprinkler

The *fire alarm riser diagram,* **Figure 41–7,** shows the new School Addition fire alarm system with ¾-inch conduits to ramp area 113. The School Addition drawing E-2 shows three ¾-inch conduits with the note "Three (3) ¾-inch existing conduits from the Fire Alarm Control Panel (150 feet)." These three conduits are to be used for the School Addition fire alarm connection to the existing fire alarm system. The original fire alarm control panel was sized to accommodate this School Addition; however, you should verify that the existing *special systems* (fire alarm, security, clock, etc.) will accommodate the additional requirements when adding to or modifying the existing system(s).

The *telephone riser diagram,* **Figure 41–8,** shows an existing 1¼-inch conduit, 150 feet in length, from ramp area 113 to the existing main telephone terminal cabinet. This conduit is to be extended to the telephone terminal

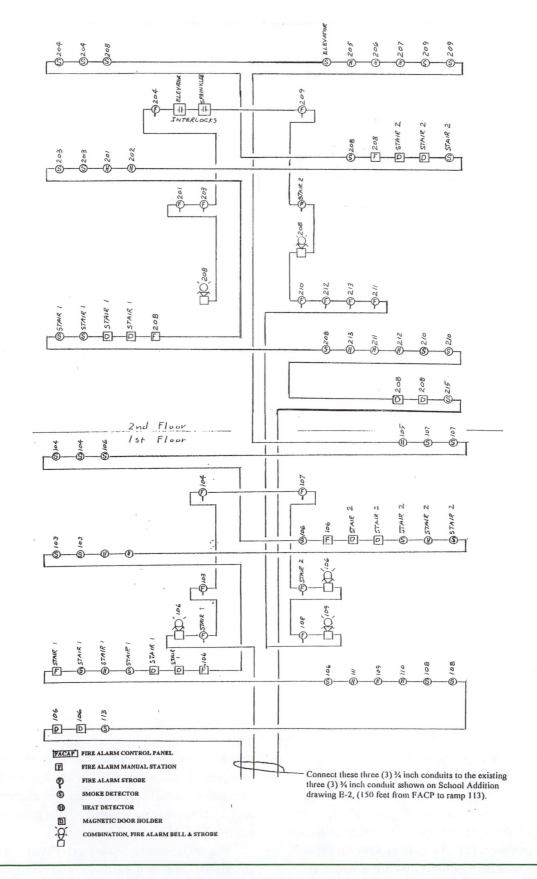

**Figure 41–7.** Fire alarm riser diagram for the School Addition.

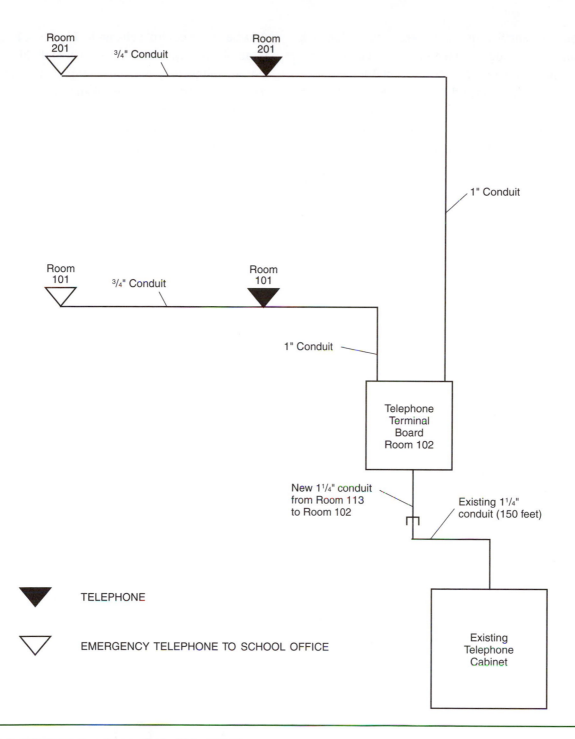

Room
201 ▽    ³/₄" Conduit    Room
201 ▼

1" Conduit

Room
101 ▽    ³/₄" Conduit    Room
101 ▼

1" Conduit

Telephone
Terminal
Board
Room 102

New 1¹/₄" conduit
from Room 113
to Room 102

Existing 1¹/₄"
conduit (150 feet)

▼    TELEPHONE

▽    EMERGENCY TELEPHONE TO SCHOOL OFFICE

Existing
Telephone
Cabinet

**Figure 41–8.** Telephone riser diagram for the School Addition.

board in room 102, First Floor Plan E-7, drawing E-2. The telephone riser diagram shows telephone conduit to be installed in the area above the suspended ceiling from the telephone terminal board to the outlet locations as shown on drawing E-2 and outlet detail A-11.

# Schematic Diagrams

A schematic wiring diagram is a drawing that uses symbols and lines to show how the parts of an electrical assembly or unit are connected. A schematic does not necessarily show where parts are actually located, but it does explain how to make electrical connections. Several of the schematics with the School Addition show connections to wires labeled G, N, and H. These stand for *ground, neutral,* and *hot.* Schematics are commonly drawn for electrical equipment that involves internal wiring—everything from washing machines to computers.

A basic motor control schematic, **Figure 41–9,** shows a three-phase power source (L1, L2, and L3) through the starter contacts (M) and the overloads (OL) powering the motor. The starter control is taken from phase L1 through a stop push button, a start push button paralleled with a latching contact "M" to the starter coil "M," and finally through the normally closed overload contacts back to

SCHEMATIC

WIRING DIAGRAM

**Figure 41–9.** Schematic diagram and wiring diagram for a three-phase, ac magnetic, nonreversing motor starter.

324   Unit 41

| MARK | MAKE | MODEL | VOLTS | WATTS | LAMP TYPE | LAMP NO./FIXT. | REMARKS |
|------|------|-------|-------|-------|-----------|----------------|---------|
| A | STONCO | VWXL11GC | 120 | 100 | 100WA19 | 1 | CAST GUARD. |
| FA | WILLIAMS | 1222-RWKA125 | 120 | 183 | F32T8/SP30 | 3 | SPLIT WIRE IN TANDEM PAIRS – WATTAGE IS FOR 2 FIXTURES W/ TOTAL OF 3 2-LAMP BALLASTS. ELECTRONIC BALLAST. |
| FA1 | WILLIAMS | 1222-RWKA125 | 120 | 92 | F32T8/SP30 | 3 | ELECTRONIC BALLAST. |
| FB | WILLIAMS | 2922-KA | 120 | 61 | F32T8/SP30 | 2 | ELECTRONIC BALLAST. |
| FC | KIRLIN | 96617-45-46-61 | 120 | 36 | PLC13W/27K | 2 | HPF BALLAST. |
| FC1 | KIRLIN | 96617-46-61-SM | 120 | 36 | PLC13W/27K | 2 | HPF BALLAST, SURFACE MOUNT; COLOR AS SELECTED |
| FD | WILLIAMS | 8222 | 120 | 61 | F32T8/SP30 | 2 | ELECTRONIC BALLAST. |
| FE | WILLIAMS | 1262-RWKA125 | 120 | 84 | F17T8/SP30 | 4 | ELECTRONIC BALLAST. |
| FF | TERON | EE26-P-H | 120 | 36 | PL13/27 | 2 | HPF BALLAST. |
| FG | WILLIAMS | 2122-IM | 120 | 61 | F32T8/SP30 | 2 | ELECTRONIC BALLAST. |
| FH | WILLIAMS | EPG-R272RWKA-125 | 120 | 61 | F32T8/SP30 | 2 | TO MATCH EXISTING |
| M1 | KIRLIN | SS-51277-24-43 (35W)-45-46-FR | 120 | 40 | 35W HPS | 1 | HPF BALLAST. |
| M2 | STONCO | PAR250LX | 120 | 300 | 250W HPS | 1 | HPF BALLAST EQUIP WITH MOUNTING BRACKET ARM. |
| EXITS | LITHONIA | WLES SERIES | 120 | 65 | LED | 1 | SEE SHEET E-2 FOR LOC. ARROWS & MOUNTING AS IND EQUIP W/ BATTERY BACK-UP |
| EMERG.LGT. BATT.PACK | EXIDE | B200 | 120/12 | 150 | H1212 | | |
| REMOTE LAMP | EXIDE | H1212 | 12 | 24 | H1212 | 2 | |

**Figure 41-10.** Fixture schedule.

phase L2. The connection label numbers (1, 2, 3, etc.), shown on the schematic, aid the electrician in troubleshooting. Several different labeling methods are used, but they all follow the same principles; and so if you understand one method, you will be able to understand the other methods.

## Schedules

An electrical schedule is used to systematically list equipment, loads, devices, and information. Schedules organize the information in an easily understood form and can be a valuable method for communicating the design requirements to contractors and their installers.

The *fixture schedule,* **Figure 41-10,** lists complete information about each fixture type shown on the lighting plan. The following is a list of the kind of information that is usually shown on a fixture schedule:

O *Mark*—The label used to indicate a fixture type. The mark is written on or next to each fixture on the plan.

O *Make*—The identification of the manufacturer being used to establish the specific design requirements needed to aesthetically and functionally light the various rooms or areas.

O *Volts*—It is becoming increasingly common for light fixtures to be powered by up to 277/480 volts, while the *switches* might use only 24 volts (low-voltage lighting control) to protect the person using the switch.

○ *Watts*—It is necessary to know the wattage of the lamps (bulbs) in each fixture, so that the space will have the amount of illumination intended by the electrical designer.

○ *Lamp Type and Quantity*—Lamp manufacturers have similar systems of designating lamp characteristics, **Figures 41–11, 41–12,** and **41–13.**

○ *Notes or Remarks*—This column is for information that does not clearly belong under the other headings.

The School Addition Fixture Schedule A-9, Floor Plans E-7 and E-14 on drawing E-1, shows type "FA" light fixture to be the light source for the classrooms. The School Addition fixture schedule, as shown in **Figure 41–10,** shows that the fixture is manufactured by Williams and the catalog number is 1222-RWKA125 with three F32T8/SP30 lamps rated at 120 volts with a 183-watt load. Also, the Remarks column indicates some special wiring and specific wiring and specific ballast-type requirements. The first and second floor Reflected Ceiling Plans, E-15 and E-16 on drawing E-7, indicate light fixture type FA to be a 2′ × 4′ light fixture mounted end to end.

Fifteen other light fixture types are shown on the fixture schedule and on the drawings and are identified by their corresponding mark or label. All are rated at 120 volts except one emergency battery-powered remote lamp, rated at 12 volts. This information is typically given in the specifications but is more readily presented by a schedule.

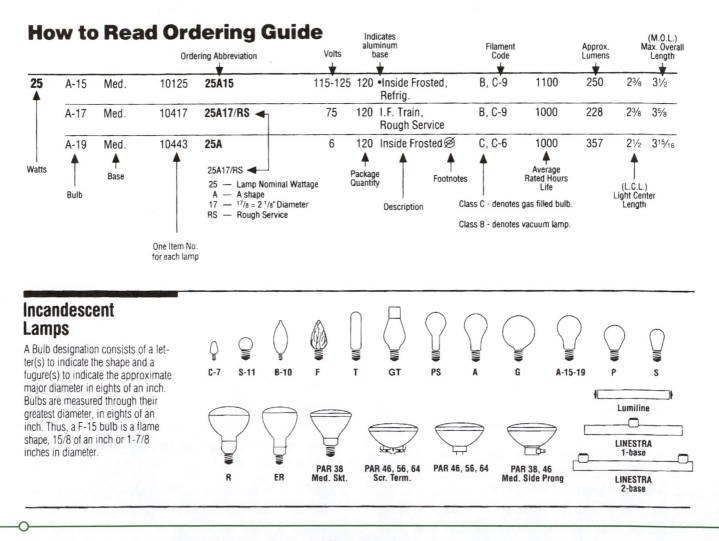

**Figure 41–11.** Incandescent lamp designations. *Courtesy of Osram Sylvania Inc.*

# How to Read Ordering Guide

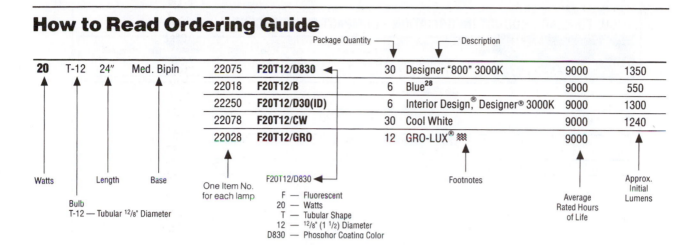

| | | | | | Package Quantity | | Description | | | |
|---|---|---|---|---|---|---|---|---|---|---|
| **20** | T-12 | 24" | Med. Bipin | 22075 | **F20T12/D830** | 30 | Designer "800" 3000K | | 9000 | 1350 |
| | | | | 22018 | **F20T12/B** | 6 | Blue[28] | | 9000 | 550 |
| | | | | 22250 | **F20T12/D30(ID)** | 6 | Interior Design,® Designer® 3000K | | 9000 | 1300 |
| | | | | 22078 | **F20T12/CW** | 30 | Cool White | | 9000 | 1240 |
| | | | | 22028 | **F20T12/GRO** | 12 | GRO-LUX® ≋ | | 9000 | |

Watts — **20**
Bulb / T-12 — Tubular 12/8" Diameter
Length — 24"
Base — Med. Bipin
One Item No. for each lamp
F20T12/D830
- F — Fluorescent
- 20 — Watts
- T — Tubular Shape
- 12 — 12/8" (1 1/2) Diameter
- D830 — Phosphor Coating Color
Footnotes
Average Rated Hours of Life
Approx. Initial Lumens

## FLUORESCENT LAMPS

The bulb shape and size of a fluorescent lamp are expressed by means of a code consisting of the letter "T" (which designates that the bulb is "tubular" in shape) followed by a number which expresses the diameter of the bulb in eighths of an inch. They vary in diameter from T-5 (5/8 inch) to T-17 (2 1/8 inches). In nominal overall length, fluorescent lamps range from 6 to 96 inches, which is always measured from back of lampholder to back of lampholder. For example, the actual overall length of the 40-watt rapid start T-12, 48 inch lamp is 47¾ inches. Circline lamps, which are circular, are available in four sizes: 6½ inches, 8 inches, 12 inches and 16 inches outside diameter. There are also U shaped fluorescent types (Curvalume™) with T-8 and T-12 bulbs. U shaped types are measured for the distance between the ends. The overall length is measured from the face of the bases to the outside of the glass bend.

T-2 Right Angle (2/8" Diameter)

**PREHEAT, RAPID START**

T-5 Miniature Bipin (5/8" diameter)

T-8 Miniature Bipin (1" diameter)

OCTRON T-8 Miniature Bipin (1" diameter)

T-12 Medium Bipin (1-1/2" diameter)

T-17 Mogul Bipin (2-1/8" diameter)

U-Shape T-12 (1-1/2" diameter)

OCTRON U-Shaped T-8 (1" diameter)

**HIGH OUTPUT AND VERY HIGH OUTPUT**

T-12 Recessed Double Contact (1-1/2" diameter)

T-14 1/2 Recessed Double Contact (1-1/4" diameter)

**SILMLINE INSTANT START**

T-6 Single Pin (3/4" diameter)

T-8 Single Pin (1" diameter)

T-12 Single Pin (1-1/2" diameter)

CIRCLINE 4-Pin T-9 (6-1/2", 8", 12", 16" outside diameters)

**Figure 41–12.** Fluorescent lamp designations. *Courtesy of Osram Sylvania Inc.*

## HOW TO READ PRODUCT INFORMATION - COMPACT FLUORESCENT

| Nominal Wattage | Bulb | MOL (in) | MOL (mm) | Base | Product Number | Ordering Abbreviation | NEMA Generic Designation | Pkg Qty | Avg Rated Life (hrs) | CCT (K) | CRI | Approx Lumens @25°C/77°F (@35°C/95°F) Initial | Mean | Symbols & Footnotes |
|---|---|---|---|---|---|---|---|---|---|---|---|---|---|---|
| 13 | Twist | 4.6 | 117 | Med | 29116 | **CF13EL/MINITWIST** | | 6 | 8000 | 3000 | 82 | 800 | 640 | ☐ ⊕ ⊕ 2,21,28, 36,63,64 |
| 26 | D(T4) | 6.8 | 173 | G24D-3 | 20710 | **CF26DD/830/ECO** | CFQ26W/G24D/30 | 50 | 10000 | 3000 | 82 | 1800 | 1548 | ▲ ☐☐2,21,28, 34,37 |
| 32 | T(T4) | 5.5 | 140 | GX24Q-3 | 20885 | **CF32DT/E/IN/835/ECO** | CFTR32W/GX24Q/35 | 50 | 10000 | 3500 | 82 | 2328 2400 | 2002 2064 | ☐2,21,28, 33,35,59 |
| 40 | L(T5) | 22.6 | 573 | 2G11 | 20586 | **FT40DL/841/RS/ECO** | FT40W/2G11/RS/41 | 10 | 20000 | 4100 | 82 | 3150 | 2709 | ☐2,21,28 |

| | |
|---|---|
| **Nominal Wattage** | Design wattage on reference ballast. Actual wattage dependent on ballast. |
| **Bulb** | Describes the shape of the bulb. |
| **Base** | Base designations for compact fluorescent lamps are the NEMA designations. Please see page 111 for base illustrations. |
| **MOL** | Maximum overall length. The actual length of the lamp measured from the bottom of the base to the top outside edge of the glass. In many cases, the bottom of the base is the bottom of the center post of the base of the lamp. |
| **Symbols & Footnotes** | Most symbols and footnotes that apply to a specific product will appear in this space. The explanations of the symbols and footnotes are at the end of the fluorescent section. |
| **Ordering Abbreviation** | A text description of the lamp. Please see below for several examples and explanations of some of the codes. |
| **NEMA Generic Designation** | Designation assigned by NEMA (National Electrical Manufacturers Association). |
| **CCT** | Correlated Color Temperature. The degree of "whiteness" of the light. Expressed in kelvins (K). Please see page 109 for more information. |
| **CRI** | Color Rendering Index. A numbering system for rating the relative color rendering quality of a light source compared to a standard. Please see page 109 for more information. |
| **Initial & Mean Lumens** | Initial lumens are measured when the lamp has been operating for 100 hours. Mean lumens are typically measured at 40% of the rated life of the lamp. Compact Fluorescent lamp lumens are measured at 25°C (77°F) and 35°C (95°F) |

### How to Read Ordering Abbreviations

| **CF26DD/830** | | **CF32DT/E/IN/835/ECO** | | **FT40DL/841/RS/ECO** | | **CF20EL/830/MED/ECO** | |
|---|---|---|---|---|---|---|---|
| CF | Compact Fluorescent | CF | Compact Fluorescent | FT | Fluorescent Twin | CF | Compact Fluorescent |
| 26 | Nominal lamp wattage | 32 | Nominal lamp wattage | 40 | Nominal lamp wattage | 20 | Nominal lamp wattage |
| DD | DULUX® Double | DT | DULUX Triple | DL | DULUX Long | EL | Electronic Lamp |
| 8 | 82 CRI | E | Electronic or dimming | 8 | 82 CRI | 8 | 82 CRI |
| 30 | 3000K CCT | | operation | 41 | 4100K CCT | 30 | 3000K CCT |
| | | IN | Amalgam | RS | Rapid Start | MED | Medium screw base |
| | | 8 | 82 CRI | ECO | ECOLOGIC | ECO | ECOLOGIC |
| | | 35 | 3500K CCT | | | | |
| | | ECO | ECOLOGIC | | | | |

**Figure 41–13.** Compact fluorescent lamp designations. *Courtesy of Osram Sylvania Inc.*

A *panel schedule* identifies the panels by their mark or label. They are shown by this same designation on the electrical floor plan. The panel schedule for the School Addition lists each of the branch circuits served by a panel, the calculated load for that branch circuit, the voltage of that circuit, and the number of poles and trip rating for each circuit breaker. A panel schedule also might include:

O Type (surface or flush)

O Panel main buss amperes, volts, and phases

O Main circuit breaker/main lugs only

O Breaker frame sizes

O Items fed and/or remarks

On the lighting and power plans, we saw that each device is connected to a branch circuit identified by a panel number (label) and circuit number. Those numbers correspond with the numbers on the panel schedule.

When a commercial project has a kitchen, you should have a *kitchen equipment schedule.* If the drawings do not have one, it is very helpful to make one for the installer. The kitchen equipment schedule should include:

O Equipment number or designation

O Description of each equipment item

O The load in horsepower or kilowatts

## COMPACT FLUORESCENT LAMPS

The overall length of DULUX® compact fluorescent lamps is measured from the bottom of the base to the outside edge of the glass. In many cases, the bottom of the base is the bottom of the center post of the base of the lamp.

DULUX S    DULUX S/E    DULUX D    DULUX D/E    DULUX T    DULUX T/E DULUX T/E/IN

DULUX L    DULUX F

DULUX EL Triple    DULUX EL TWIST    DULUX EL CLASSIC (A-Shape)    DULUX EL BULLET    DULUX EL Low Profile GLOBE    DULUX EL GLOBE    DULUX EL DECOR

DULUX EL REFLECTOR    BR-30    PAR-38    DULUX EL Circline (6-1/2" &8" outside diameters)    BR40 Med    PAR38 Med Skrt

**Figure 41–13.** (continued)

- ○ Volts
- ○ Wire size
- ○ Conduit size
- ○ Protection in amperes
- ○ Who will furnish each equipment item (furnished by others or by contractor)

- ○ Installation requirements
- ○ Remarks column for any detailed specific information required

A *receptacle schedule* is valuable when a number of special or specific receptacle types are found on the electrical drawings. If one is not provided, you should make one to expedite the installation time required and reduce the chance of installation error. A receptacle schedule should include:

O The symbol designation used by the designer

O Amperage rating

O Number of wires and poles

O Voltage rating

O NEMA type

O The configuration of the blades or slots

O A manufacturer catalog number reference

O Special information (duplex, single, three phase, etc.)

Note:
Receptacle information may be found on an *equipment schedule.*

The local plans review and the utility company typically require a *connected load schedule.* This type of schedule includes:

O Type of load

O Building or area designation

O Size of load (kilowatts or horsepower)

O Total electrical load by type or area

O Notes explaining any special methods used in the load calculations

## Specifications

The drawings and the **specifications** are the items that establish the intended design and the construction requirements required by the owner, architect, and engineers. The contractors and their installers must review the **contract** documents for conflicts and/or discrepancies between the contract, the specifications, and the drawings. The drawings should be reviewed for conflicts between sections (architectural, mechanical, plumbing, structural, civil, HVAC, electrical, etc.). The drawings in your specific section should be reviewed for conflicts sheet by sheet.

The electrical specifications give the quality of materials intended to be used and the installation and testing requirements. Sometimes a specific manufacturer or catalog number is specified with no substitution or "equal" allowed. This usually inflates the cost of the project. There is no standard specification. The contractors and their installers should read through all project specifications and become knowledgeable of their content prior to starting any installation.

## ✓ CHECK YOUR PROGRESS

Can you perform these tasks?

● Explain where light fixtures are to be placed and how they are connected.

● Determine what branch circuit in a panel supplies a particular light fixture or electrical outlet.

● Understand and use the information on a fixture schedule.

● Use a schematic diagram to explain how an electrical device is to be wired.

## ASSIGNMENT

Refer to the drawings of the School Addition (included in the packet) to complete this assignment.

1. What are the four basic methods of showing electrical circuits?

2. How is a home run circuit shown?

3. What is a reflected ceiling plan?

4. In a low-voltage lighting system, what does the light switch activate?

5. Where do you find the industry standardized electrical symbols?

6. What are the four basic types of electrical drawings?

7. What is the name of the method used to show the path of wiring or raceway from one level of a building to another?

8. What is a schematic wiring diagram?

9. What is used to systematically list equipment, loads, devices, and information?

10. Where do you find the quality of material intended to be used on a project?

11. How many F17T8/SP30 lamps are required on the first floor of the addition?

12. Explain why one of the switches in room 109 is listed as $S_{3M}$ and the other is simply $S_3$.

13. How are the lights turned on and off in the first floor corridor?

14. What circuit carries the lights for stair #2?

15. What is the approximate total wattage of the lamps in the boys' toilet room on the first floor?

16. What is the total load for the circuit that serves the lights in the boys' toilet room on the first floor?

17. What is indicated by the D in a square near the doors from the existing building into the addition?

18. What is on circuit L10, 24?

19. Where are the devices on circuit L10, 15?

20. Explain what each of the colored terminals on a classroom lighting control is to be connected to:

Green                     ____
Orange                    ____
Black                     ____
White                     ____
Blue (inner terminal)     ____
Blue (outer terminal)     ____

# PART IV

# TEST

**A. Which of the symbols shown in Column II is used on the School Addition drawings to represent the objects or materials listed in Column I?**

| I | | II |
|---|---|---|
| 1. Room number | a. | ⊠ |
| 2. Building section | b. | ⬡ D |
| 3. Rigid insulation | c. | ○ 4 |
| 4. Window type | d. | ▽ J-6 / A201 |
| 5. Door number | e. | 209 |
| 6. Interior elevation | f. | ■── A-8 / A502 |
| 7. Wood blocking | g. | ▨ |
| 8. Detail key | h. | ◢ G14/A610 |

**B. Refer to the door schedule on School Addition Sheet A001 to write the door numbers from Column II that are associated with the material in Column I.**

| I | | II |
|---|---|---|
| 9. Wood | | 216 |
| 10. Steel hollow metal | | 104 |
| 11. Plastic | | 101 |

**C. Name the material for each of the numbered items in the drawing of a bond beam.**

12. ____

13. ____

14. ____

15. ____

16. ____

**D. Describe the material at each of the following locations in the School Addition.**

17. Door frame 103

18. South wall of corridor 106

19. Structural portion of west wall of classroom 104

20. Exterior surface of west wall at corners

21. Exterior surface of west wall below windows

22. Stair treads

23. Vertical reinforcement in elevator pit walls

24. Beam at the top of the west wall of classroom 204

25. Lintel over the windows in the west wall

26. Door frame at entrance to stair #1

**E. Refer to the School Addition drawings and write the elevation at the following locations.**

27. Second finish floor

28. Top of roof

29. Top of bricks beneath window in classroom 103

30. Top of concrete blocks in first floor corridor walls

31. Top of tackboard on north wall of classroom 103

32. Finish floor in elevator pit

33. Top of foundation wall at typical locations

34. Top of footing for column F3

35. Top of beams in second floor framing between columns in row C

**F. Refer to the School Addition drawings to answer the following questions.**

36. What is the height of the rough opening for the grille that allows outside air to enter classroom 103?

37. What supports the outer few inches of the roofing membrane in the area where the roof-edge coping meets it?

38. What size are the concrete masonry units in the north wall?

39. What are the three materials that make up the roof decking on the canopy over door 101?

40. What are the drawing and sheet numbers for information about typical column base plates?

41. Give as much information as possible about the welds that fasten the roof framing bar joists to the beams at the tops of the north and south walls.

42. On what branch circuit are the light fixtures in stair #1?

43. How are the light fixtures in the second floor corridor turned on and off?

44. How many lamps are required for all the fixtures in the second floor corridor?

45. What branch circuit supplies the convenience outlets in the second floor corridor?

46. Which terminal on the corridor lighting control unit is connected to the neutral leg of the supply?

47. What is the diameter of the sanitary drain where it leaves the building?

48. What are the pipe sizes that storm water passes through as it flows from the roof drain on the north canopy to the point where it exits the building?

49. Where is the nearest cleanout to the roof drain at the west end of the addition?

50. What type of unit provides heat in storage room 109?

# SCHOOL ADDITION MASTER KEYNOTES

# Master Keynotes

2.11   Demolition, Removals & Relocation

2.20  Site Preparation & Earthwork
  2.20A    Select Fill - Bank Run Gravel
  2.20B    Select Granular Material
  2.20D    Topsoil
  2.20F    Crushed Gravel
  2.20G    4" Perforated P.V.C.
  2.20I    Compacted Subgrade

2.22  Structural Excavation, Backfill and Compaction (Building Area)

2.60  Pavement and Walks
  2.60A    Vehicular Area Sub-Base Course Granular
  2.60B    Asphaltic Concrete - Binder Course
  2.60C    Asphaltic Concrete Top - Wearing Course
  2.60D    Concrete Walk/Paving
  2.60E    Precast Concrete Curb
  2.60G    Reinforcing Mesh
  2.60H    Expansion Joint Filler
  2.60I    Control Joint - 4'-0" O.C. Maximum Saw Cut or Tooled
  2.60J    Stabilization Fabric
  2.60P    Asphaltic Concrete-Base Course
  2.60Q    Detectable Warning Pavers

2.61  Pavement Markings
  2.61B    Painted ANSI Handicap Symbol
  2.61C    Painted Traffic Control Lines

2.80  Landscaping
  2.80A    Seeding

2.83  Chain Link Fence
  2.83D    2.875" O.C. Corner Post
  2.83E    2.875" O.C. Line Post
  2.83F    1-1/2" Top Rail
  2.83G    2" Mesh Fabric
  2.83H    7 Gauge Tension Wire Continuous
  2.83I    Post Cap
  2.83J    11 Gauge Rail Clamps
  2.83K    Sleeve
  2.83L    11 Gauge Bands
  2.83M    1/4" X 3/4" Turnbuckle
  2.83N    Stretcher Rod
  2.83O    3/8" Diameter Truss Rod

2.83P    Pass Thru Cap
2.83Q    Gate Frame
2.83R    Top Hinge
2.83S    Hinge
2.83T    Fork Latch

3.30 Concrete
See Structural Drawings*
3.30A    Cast-In-Place-Concrete
3.30B    Reinforcing Bar
3.30D    Welded Wire Mesh, Size as Noted

3.32 Concrete Slab on Grade

3.325 Concrete Slab on Metal Deck

3.40 Precast Concrete
3.40D    Precast Dry Pit
3.40E    Precast Concrete Wall Unit
3.40F    Precast Concrete Cap Unit
3.40G    Geogrid

4.10 Mortar and Masonry Grout
4.10A    Grout Solid
4.10B    Mortar Fill

4.20 Unit Masonry
4.20A    Face Brick - Standard Modular
4.20C    Concrete Masonry Units
4.20D    Horizontal Joint Reinforcing
4.20E    Flexible Masonry Anchors
4.20G    Premolded Control Joint Strips
4.20I    Reinforcing Bars
4.20J    Flexible Masonry Column Tie Anchors
4.20K    Weep Holes
4.20M    Control Joint
4.20P    Slate Sill - 1/4" Chamfer at Exposed Edges
4.20S    Precast Concrete Sill
4.20Y    Precast Concrete Band
4.20AA   Expansion Joint Closure

5.10 Structural Steel
See Structural Drawings*
5.10A    Steel Beam
5.10B    Steel Angle
5.10D    Anchor Bolt
5.10E    Steel Tube Column
5.10F    Miscellaneous Steel - Shape and Size as Noted

5.10G   Steel Tee

5.20 Steel Joists
See Structural Drawings*

5.30 Metal Decking
See Structural Drawings*
5.30A   Metal Roof Decking

5.50 Miscellaneous Metals
5.50A   Miscellaneous Steel - Shape and Size as Noted
5.50B   Steel Angle
5.50C   1-1/4" Nominal Steel Pipe (1.66" O.D.) Post
        4'-0" O.C. Maximum
5.50D   10 Gauge Metal Riser, Tread and Landing Pan
5.50E   MC 12 X 10.6 Stringer
5.50F   3-1/2" X 2-1/2" X 1/4" - LLV - Pan Support at 3"-0" O.C.
        Maximum
5.50G   1-1/2" X 1-1/2" X 1/4" Steel Angle Carriers
5.50M   Galvanized Steel Bent Plate
5.50N   Galvanized Steel Flat Bar Stock
5.50O   3/4" Diameter Galvanized Steel Rungs
5.50S   Steel Bollards - Fill With Concrete - Bevel Concrete at Top
5.50U   1-1/4" Nominal Steel Pipe (1.66" O.D.)
5.50Y   Countersunk Machine Head Screw/Expansion
5.50A   F Steel Plate
5.50A   M 1" Nominal Steel Pipe Handrail (1.315" O.D.)
5.50A   N 1" Nominal Steel Pipe Bracket
5.50A   P Steel Wall Bracket, Julius Blum #622, or Equal
5.50A   X Aluminum Sleeve (.050")
5.50A   Y Steel Plate, Size as Noted

5.80 Expansion Joints
5.80A   Expansion Joint Cover

6.10 Rough Carpentry
6.10A   Wood Blocking
6.10B   Plywood
6.10C   Wood Furring
6.10I   Wood Framing 16" O.C.
6.10S   Joist Hanger
6.10T   Pressure Treated Rail; Size as Noted
6.10U   Pressure Treated Post; Size as Noted

6.20 Finish Carpentry
6.20F   Softwood Trim

7.15 Dampproofing

7.20  Insulation
  7.20B     Batt Insulation
  7.20C     Cavity Wall Insulation
  7.20H     Metal Z Furring

7.41  Preformed Wall Panels
  7.41C     Vertical Siding
  7.41F     Trim to Match Siding

7.50  Roofing System (EPDM)
  7.50A     Roofing Membrane
  7.50B     Membrane Flashing
  7.50C     Insulation
  7.50D     Thermal Barrier
  7.50E     Tapered Insulation
  7.50F     Vapor Barrier/Air Seal
  7.50G     Ballast and Protective Mat
  7.50H     Fascia System
  7.50K     Fascia Extender
  7.50M     Aluminum Counterflashing
  7.50N     Sealant
  7.50P     Termination Bar
  7.50Q     Expansion Joint Support
  7.50R     Closed Cell Backer Rod
  7.50T     Insert Drain
  7.50V     Recovery Board
  7.50W     Aluminum Trim - .032" Thickness

7.51  Roofing System (Silicone)
  7.51A     Silicone Coating
  7.51B     Urethane Insulation
  7.51D     Sealant
  7.51E     Aluminum Foam Stop - 0.050" Thickness
  7.51J     Aluminum Trim - 0.032" Thickness
  7.51K     Insert Drain

7.53  Roofing System (EPDM-Foam Adhesive)
  7.53A     Roofing Membrane
  7.53B     Foam Adhesive
  7.53C     Membrane Flashing
  7.53D     Insulation
  7.53H     Fascia System
  7.53K     Fascia Extender
  7.53O     Sealant
  7.53Q     Termination Bar
  7.53R     Closed Cell Backer Rod
  7.53T     Insert Drain

7.60 Flashing and Sheetmetal
  7.60A    Thru-wall Flashing - Turn-Up at Ends at Wall Openings to Form Dam

7.72 Roof Accessories

7.90 Caulking and Sealants
  7.90A    Sealant and Backing Material (Size as Required to Fill Void)
  7.90B    Foam Joint Filler - Width Shown X Size Required to Fill Void)
  7.90C    Sealant Bead
  7.90G    Expansion Joint Filler

8.10 Metal Doors and Frames
  8.10A    Steel Hollow Metal Door Frame
  8.10B    Steel Hollow Metal Door
  8.10E    Metal Access Door
  8.10F    16 Gauge Hollow Metal Pipe Enclosure

8.15 Plastic Doors
See Door Schedule*
  8.15A    Plastic Door
  8.15B    Aluminum Door Frame System
  8.15C    .090" Break Metal to Match Door Frame System
  8.15G    Aluminum Angle (Size as Noted)

8.20 Wood Doors
See Door Schedule*
  8.20A    Wood Door

8.33 Rolling Counter Fire Door
  8.33B    Vertical Sliding Pass Window
  8.33C    Stainless Steel Sill

8.50 Metal Windows
  8.50A    Aluminum Windows
  8.50B    1/8" Break Metal to Match Windows
  8.50C    Extruded Aluminum Sill and Clip
  8.50D    Extruded Aluminum Head/Jamb Panning
  8.50E    Extruded Aluminum Trim
  8.50K    Fixed Aluminum Window
  8.50L    Horizontal Rolling Aluminum Window
  8.50R    Foam Tape
  8.50S    Foam Insulation
  8.50T    Double Hung Window

8.70 Hardware and Specialties
See Door Schedule*
  8.70D    Aluminum Threshold Set in Mastic

8.80 Glazing
  8.80A    Tempered Insulating Glass
  8.80B    Insulating Glass
  8.80C    Safety Wire Glass
  8.80D    Tempered Glass
  8.80I    Muntins Inside Insulated Glass
  8.80K    Obscure Insulating Glass

09805    Encapsulation of Asbestos Containing Material

9.25 Gypsum Wallboard
  9.25A    5/8" Type 'X' Gypsum Wallboard
  9.25B    "J" Casing Bead
  9.25C    Metal Stud System
  9.25D    Metal Furring
  9.25G    Metal Stud Runner - 2" Deep to Provide 1" Expansion for Studs
  9.25J    Gypsum Sheathing
  9.25k    Carrying Channel
  9.25L    5/8" Exterior Gypsum Ceiling Board
  9.25N    Metal Angle Runner

9.30 Tile
See Room Finish Schedule*
  9.30A    Glazed Wall Tile
  9.30B    Unglazed Ceramic Mosaic Tile
  9.30D    Glazed Ceramic Tile Wall Base
  9.30E    Marble Thresholds - See Door Schedule*
  9.30F    Accent Tile - Continuous Around Room
  9.30G    Unglazed Ceramic Mosaic Tile Base

9.40 Terrazzo
  9.40A    Thin Set Terrazzo
  9.40C    Terrazzo Cove Base

9.50 Acoustical Treatment
See Room Finish Schedule*
  9.50A    Suspended Ceiling System

9.65 Resilient Flooring
See Room Finish Schedule*
  9.65A    Vinyl Composition Tile
  9.65B    Vinyl Cove Base
  9.65D    Rubber Stair Treads and Risers
  9.65E    Molded Rubber Tile
  9.65J    Rubber Base

9.80 Special Coating System
  9.80A    Special Coating System

9.80B    Finish Coat
9.80C    Base Coat
9.80D    Reinforcing Mesh
9.80E    Insulation
9.80F    Gypsum Sheathing
9.80G    Metal Studs
9.80H    Sealant and Backer Rod
9.80I    Waterproof Base Coat
9.80J    Routed Joint
9.80K    Insulation Board Below Grade
9.80L    Below Grade Waterproofing
9.80M    Slide Clip
9.80N    Bent Galvanized Metal (Size and Gauge as Noted)
9.80O    Expansion Joint

9.90 Painting
See Room Finish Schedule*
9.90E-1 Paint E-1
9.90E-2 Paint E-2
9.90P-1 Paint P-1
9.90P-2 Paint P-2
9.90P-3 Paint P-3
9.90P-4 Paint P-4
9.90P-6 Paint P-6
9.90P-8 Paint P-8

10.10 Chalkboards and Tackboards
10.10B Liquid Marker Board
10.10C Tackboard
10.10F Projection Screen

10.25 Firefighting Devices
10.25A Fire Extinguisher Cabinet. Paint P-2. Model #2409-R2 By Larson
           Manufacturing Co., or Approved Equal

10.42 Signage and Graphics
10.42A 12" X 18" Aluminum Handicap Sign

10.50 Lockers
10.50A Single Tier Lockers
10.50G Sloping Top
10.50H Metal Base
10.50I  Recessed Trim

10.62 Folding Partitions
10.62.A Folding Partitions
10.62B Tracks, Support Brackets, Hangar Rods and Finish Trim
10.62D Liquid Marker Board

10.62E Tackboard

10.80 Toilet Accessories
10.80C Mirror
10.80D Grab Bars

11.46 Unit Kitchens
11.46A Compact Kitchen Unit

12.17 Entrance Mats
12.17A Recessed Entrance Mat

12.30 Casework
See Drawings for Casework
12.30A Plastic Laminate Counter with 4" Backsplash
12.30D Filler Strip - Size as Required
12.30G Base Cabinet
12.30I Lab Countertop
12.30K Plastic Laminate Top
12.30L Wall Cabinet

14.20 Elevators
14.20A Elevator

*See reference noted for information elaborating on or work in addition to materials noted here. Provide all materials as required for a complete and proper installation.

# Glossary

**Addendum** A change or modification to the bid documents, plans, and specifications that is made prior to the contractor's bid date

**Aggregate** Hard materials such as sand and crushed stone used to make concrete

**Ampere (amp)** Unit of measure of electric current

**Anchor Bolt** A bolt placed in the surface of concrete for attaching wood framing members

**Apron** Concrete slab at the approach to a garage door—also the wood trim below a window stool

**Architect's Scale** A flat or triangular scale used to measure scale drawings

**Ash Dump** A small metal door in the bottom of a fireplace

**Awning Window** A window that is hinged near the top so the bottom opens outward

**Backfill** Earth placed against a building wall after the foundation is in place

**Backsplash** The raised lip on the back edge of a countertop to prevent water from running down the backs of the cabinets

**Balloon Framing** Type of construction in which the studs are continuous from the sill to the top of the wall—upper floor joists are supported by a let-in ribbon

**Balusters** Vertical pieces that support a handrail

**Batt Insulation** Flexible, blanket-like pieces, usually of fiberglass, used for thermal or sound insulation

**Batten** Narrow strip of wood used to cover joints between boards of sheet materials

**Batter Boards** An arrangement of stakes and horizontal pieces used to attach lines for laying out a building

**Beam** Any major horizontal structural member

**Beam Pocket** A recessed area to hold the end of a beam in a concrete or masonry wall

**Board Foot** One hundred forty-four cubic inches of wood or the amount contained in a piece measuring 12" × 12" × 1"

**Bottom Chord** The bottom horizontal member in a truss

**Box Sill** The header joist nailed across the ends of floor joists at the sill

**Branch Circuit** The electrical circuit that carries current from the distribution panel to the various parts of the building

**British Thermal Unit (BTU)** The amount of heat required to raise the temperature of 1 pound of water 1 degree Fahrenheit

**Building Lines** The outside edge of the exterior walls of a building

**Casement Window** A window that is hinged at one side so the opposite side opens outward

**Casing** The trim around a door or window

**Centerline** An actual or imaginary line through the exact center of any object

**Change Order** A change or modification to the contract documents, plans, and specifications that is made after the contract has been awarded to the selected trade contractor

**Cleanout** A pipe fitting with a removable plug that allows for cleaning the run of piping in which it is installed, or an access door at the bottom of a chimney

**Collar Beam** Horizontal members that tie opposing rafters together, usually installed about halfway up the rafters

**Column** A metal post to support an object above

**Common Rafter** A rafter extending from the top of the wall to the ridge

**Concrete** Building material consisting of fine and coarse aggregates bonded together by portland cement

**Conductor** Electrical wire—a cable may contain several conductors

**Contour Lines** Lines on a topographic map or site plan to describe the contour of the land

**Contract** Any agreement in writing for one party to perform certain work and the other party to pay for the work

**Convenience Outlet** Electrical outlet provided for convenient use of lamps, appliances, and other electrical equipment

**Cornice** The construction that encloses the ends of the rafters at the top of the wall

**Cornice Return** The construction where the level cornice meets the sloping rake cornice

**Course** A single row of building units such as concrete blocks or shingles

**Cove Mold** Concave molding used to trim an inside corner

**Damper** A door installed in the throat of a fireplace to regulate the draft

**Dampproofing** Vapor barrier or coating on foundation walls or under concrete slabs to prevent moisture from entering the house

**Datum** A reference point from which elevations are measured

**Delivery Sheet (Trusses)** A summary sheet included with a packet of truss drawings to show how many of each type of truss is required

**Detail** A drawing showing special information about a particular part of the construction—details are usually drawn to a larger scale than other drawings and are sometimes section views

**Dormer** A raised section in a roof to provide extra headroom below

**Double-hung Window** A window consisting of two sashes that slide up and down past one another

**Drip Cap** A wood ledge over wall openings to prevent water from running back under the frame or trim around the opening

**Drip Edge** Metal trim installed at the edge of a roof to stop water from running back under the edge of the roof deck

**Drywall** Interior wall construction using gypsum wallboard

**Elevation** A drawing that shows vertical dimensions—it may also be the height of a point, usually in feet above sea level

**Fascia** The part of a cornice that covers the ends of the rafters

**Firestop** Blocking or noncombustible material between wall studs to prevent vertical draft and flame spread

**Flashing** Sheet metal used to cover openings and joints in walls and roofs

**Float** To level concrete before it begins to cure—floating is done with a tool called a float

**Floor Plan** A drawing showing the arrangement of rooms, the locations of windows and doors, and complete dimensions—a floor plan is actually a horizontal section through the entire building

**Flue** The opening inside a chimney—the flue is usually formed by a terra cotta flue liner

**Flush Door** A door having flat surfaces

**Footing** The concrete base upon which the foundation walls are built

**Footing Drain** (Also called Perimeter Drain) An underground drain pipe around the footings to carry groundwater away from the building

**Frieze** A horizontal board beneath the cornice and against the wall above the siding

**Frostline** The maximum depth to which frost penetrates the earth

**Furring** Narrow strips of wood attached to a surface for the purpose of creating a plumb or level surface for attaching the wall, ceiling, or floor surface

**Gable** The triangular area between the roof and the top plate walls at the ends of a gable roof

**Gable Studs** The studs placed between the end rafters and the top plates of the end walls

**Gauge** A standard unit of measurement for the diameter of wire or the thickness of sheet metal

**Girder** A beam that supports floor joists

**Grout** A thin mixture of high-strength concrete or mortar

**Gypsum Wallboard** Drywall material made of gypsum encased in paper to form boards

**Header** A joist fastened across the ends of regular joists in an opening, or the framing member above a window or door opening

**Hearth** Concrete or masonry apron in front of a fireplace

**Hip** Outside corner formed by intersecting roofs

**Hip Rafter** The rafter extending from the corner of a building to the ridge at a hip

**Hose Bibb** An outside faucet to which a hose can be attached

**Insulated Glazing** Two or more pieces of glass in a single sash with air space between them for the purpose of insulation

**Isometric Drawing** A kind of drawing in which horizontal lines are 30 degrees from true horizontal and vertical lines are vertical

**Jack Rafter** Rafter between the outside wall and a hip rafter or the ridge and a valley rafter

**Jamb** Side members of a door or window frame

**Joists** Horizontal framing members that support a floor or ceiling

**Lintel** Steel or concrete member that spans a clear opening—usually found over doors, windows, and fireplace openings

**Masonry Cement** Cement that is specially prepared for making mortar

**Mil** A unit of measure for the thickness of very thin sheets—1 mil equals .001"

**Miter** A 45-degree cut so that two pieces will form a 90-degree corner

**Mortar** Cement and aggregate mixture for bonding masonry units together

**Mullion** The vertical piece between two windows that are installed side by side—window units that include a mullion are called mullion windows

**Muntins** Small vertical and horizontal strips that separate the individual panes of glass in a window sash

**Nailer** A piece of wood used in any of several places to provide a nailing surface for other framing members

**Nominal Size** The size by which a material is specified—the actual size is often slightly smaller

**Nosing** The portion of a stair tread that projects beyond the riser

**Orthographic Projection** A method of drawing that shows separate views of an object

**Panel Door** A door made up of panels held in place by rails and stiles

**Penny Size** The length of nails

**Perimeter Drain** (See Footing Drain)

**Pilaster** A masonry or concrete pier built as an integral part of a wall

**Pitch** Refers to the steepness of a roof—the pitch is written as a fraction with the rise over the span

**Plan View** A drawing that shows the layout of an object as viewed from above

**Plate** The horizontal framing members at the top and bottom of the wall studs

**Platform Framing** (Also called Western Framing) A method of framing in which each level is framed separately—the subfloor is laid for each floor before the walls above it are formed

**Plenum** A chamber within a forced-air heating system that is pressurized with warm air

**Plumb** Truly vertical or true according to a plumb bob

**Ply (Trusses)** If trusses are plied, they are joined together face-to-face to make a stronger unit

**Potable Water** Water that is safe for drinking

**Portland Cement** Finely powdered limestone material used to bond the aggregates together in concrete and mortar

**R value** The ability of a material to resist the flow of heat

**Rafter** The framing members in a roof

**Rail** The horizontal members in a door, sash, or other panel construction

**Rake** The sloping cornice at the end of a gable roof

**Resilient Flooring** Vinyl, vinyl-asbestos, and other man-made floor coverings that are flexible yet produce a smooth surface

**Ridge Board** The framing member between the tops of rafters that runs the length of the ridge of a roof

**Rise** The vertical dimension of a roof or stair

**Riser** The vertical dimension of one step in a stair—the board enclosing the space between two treads is called a riser

**Rowlock** Position of bricks in which the bricks are laid on edge

**Run** The horizontal distance covered by an inclined surface such as a rafter or stair

**Sash** The frame holding the glass in a window

**Saturated Felt** Paperlike felt that has been treated with asphalt to make it water resistant

**Screed** A straight board used to level concrete immediately after it is placed

**Section View** A drawing showing what would be seen by cutting through a building or part

**Setback** The distance from a street or front property line to the front of a building

**Sheathing** The rough exterior covering over the framing members of a building

**Shim** Thin material, typically wood shingle, used to adjust a small space

**Sill** The framing member in contact with a masonry or concrete foundation

**Sill Sealer** Compressible material used under the sill to seal any gaps

**Site Constructed** Built on the job

**Site Plan** The drawing that shows the boundaries of the building, its location, and site utilities

**Sliding Window** A window with two or more sashes that slide horizontally past one another

**Soffit** The bottom surface of any part of a building, such as the underside of a cornice or lowered portion of a ceiling over wall cabinets

**Soldier** Brick position in which the bricks are stood on end

**Span** The horizontal dimension between vertical supports—the span of a beam is the distance between the posts that support it

**Specifications** Written description of materials or construction

**Square** The amount of siding or roofing materials required to cover 100 square feet

**Stack** The main vertical pipe into which plumbing fixtures drain

**Stair Carriage** The supporting framework under a stair

**Stile** The vertical members in a sash, door, or other panel construction

**Stool** Trim piece that forms the finished window sill

**Stop** Molding that stops a door from swinging through the opening as it is closed—also used to hold the sash in place in a window frame

**Stud** Vertical framing member in a wall

**Subfloor** The first layer of rough flooring applied to the floor joists

**Sweat** Method of soldering used in plumbing

**Termite Shield** Sheet metal shield installed at the top of a foundation to prevent termites from entering the wood superstructure

**Thermal-break Window** Window with a metal frame that has the interior and exterior separated by a material with a higher R value than the metal itself.

**Thermostat** An electrical switch that is activated by changes in temperature

**Top Chord** The top horizontal member of a truss

**Trap** A plumbing fitting that holds enough water to prevent sewer gas from entering the building

**Tread** The surface of a step in stair construction

**Trimmers** The double framing members at the sides of an opening

**Truss** A manufactured assembly used to support a load over a long span

**Truss Detail** An engineering drawing giving detailed specification for the manufacture of one particular truss. A truss drawing packet will usually include a truss detail for each type of truss to be used.

**Truss Drawings** Any of several types of drawings provided by an engineer to give detailed information about the construction and placement of trusses

**Truss Layout Plan** A plan drawing giving detailed information about where each truss is to be placed, including spacing between trusses

**Underlayment** Any material installed over the subfloor to provide a smooth surface over which floor covering will be installed

**Valley** The inside corner formed by intersecting roofs

**Valley Rafter** The rafter extending from an inside corner in the walls to the ridge at a valley

**Vapor Barrier** Sheet material used to prevent water vapor from passing through a building surface

**Veneer** A thin covering; in masonry, a single wythe of finished masonry over a wall; in woodwork, a thin layer of wood

**Vent Pipe** A pipe, usually through the roof, that allows atmospheric pressure into the drainage system

**Vertical Contour Interval** The difference in elevation between adjacent contour lines on a topographic map or site plan

**Volt** The unit of measurement for electrical force

**Water Closet** A plumbing fixture commonly called a toilet

**Watt** The unit of measurement of electrical power—1 watt is the amount of power from 1 ampere of current with 1 volt of force

**Weep Hole** A small hole through a masonry wall to allow water to pass

**Western Framing** (See Platform Framing)

**Wythe** A single thickness of masonry construction

# Math Reviews

## MATH REVIEWS

# MATH REVIEW 1
## Fractions and Mixed Numbers—
## Meanings and Definitions

○ A *fraction* is a value that shows the number of equal parts taken from a whole quantity. A fraction consists of a numerator and a denominator.

$\frac{7}{16}$ ←Numerator
←Denominator

○ *Equivalent fractions* are fractions that have the same value. The value of a fraction is not changed by multiplying the numerator and denominator by the same number.

─────────○

**Example** Express $\frac{5}{8}$ as thirty-seconds.
Determine what number the denominator is multiplied by to get the desired denominator. $(32 \div 8 = 4)$
Multiply the numerator and denominator by 4.

$\frac{5}{8} = \frac{?}{32}$

$\frac{5}{8} \times \frac{4}{4} = \frac{20}{32}$

●─────────

○ The *lowest common denominator* of two or more fractions is the smallest denominator that is evenly divisible by each of the denominators of the fractions.

─────────○

**Example 1** The lowest common denominator of, $\frac{3}{4}, \frac{5}{8}, \frac{13}{32}$ and is 32, because is the smallest number evenly divisible by 4, 8, and 32.

$32 \div 4 = 8$
$32 \div 8 = 4$
$32 \div 32 = 1$

●─────────

─────────○

**Example 2** The lowest common denominator of, $\frac{2}{3}, \frac{1}{5}, \frac{7}{10}$ and is 30, because is 30 is the smallest number evenly divisible by 3, 5, and 10.

$30 \div 3 = 10$
$30 \div 5 = 6$
$30 \div 10 = 3$

●─────────

○ *Factors* are numbers used in multiplying. For example, 3 and 5 are factors of 15.

$3 \times 5 = 15$

○ A fraction is in its *lowest terms* when the numerator and the denominator **do not** contain a common factor.

─────────○

**Example** Express $\frac{12}{16}$ in lowest terms.
Determine the largest common factor in the numerator and denominator. The numerator and the denominator can be evenly divided by 4.

$\frac{12 \div 4}{16 \div 4} = \frac{3}{4}$

●─────────

○ A *mixed number* is a whole number plus a fraction.

$6 \ \frac{15}{16}$

Whole Number ──↑  ↑── Fraction

$6 + \frac{15}{16} = 6\frac{15}{16}$

○ *Expressing fractions as mixed numbers.* In certain fractions, the numerator is larger than the denominator. To express the fraction as a mixed number, divide the numerator by the denominator. Express the fractional part in lowest terms.

──────────────○

**Example** Express $\frac{38}{16}$ as a mixed number.

Divide the numerator 38 by the denominator 16.

Express the fractional part $\frac{6}{16}$ in lowest terms.

Combine the whole number and fraction.

$$\frac{38}{16} = 2\frac{6}{16}$$

$$\frac{6 \div 2}{16 \div 2} = \frac{3}{8}$$

$$\frac{38}{16} = 2\frac{3}{8}$$

●───────────────

○ *Expressing mixed numbers as fractions.* To express a mixed number as a fraction, multiply the whole number by the denominator of the fractional part. Add the numerator of the fractional part. The sum is the numerator of the fraction. The denominator is the same as the denominator of the original fractional part.

──────────────○

**Example** Express $7\frac{3}{4}$ as a fraction.

Multiply the whole number 7 by the denominator 4 of the fractional part ($7 \times 4 = 28$). Add the numerator 3 of the fractional part to 28. The sum 31 is the numerator of the fraction. The denominator 4 is the same as the denominator of the original fractional part.

$$\frac{7 \times 4 + 3}{4} = \frac{31}{4}$$

or

$$\frac{7}{1} \times \frac{4}{4} = \frac{28}{4}$$

$$\frac{28}{4} + \frac{3}{4} = \frac{31}{4}$$

●───────────────

# MATH REVIEW 2
## Adding Fractions

○ Fractions must have a common denominator in order to be added.

○ To add fractions, express the fractions as equivalent fractions having the lowest common denominator. Add the numerators and write their sum over the lowest common denominator. Express the fraction in lowest terms.

──────────────○

**Example** Add: $\frac{3}{8} + \frac{3}{4} + \frac{3}{16} + \frac{1}{32}$

Express the fractions as equivalent fractions with 32 as the denominator.

Add the numerators.

$$\frac{3}{8} = \frac{3}{8} \times \frac{4}{4} = \frac{12}{32}$$

$$\frac{1}{4} = \frac{1}{4} \times \frac{8}{8} = \frac{8}{32}$$

$$\frac{3}{16} = \frac{3}{16} \times \frac{2}{2} = \frac{6}{32}$$

$$+\frac{1}{32} = \qquad \frac{1}{32}$$

$$\frac{27}{32}$$

●───────────────

○ After fractions are added, if the numerator is greater than the denominator, the fraction should be expressed as a mixed number.

**Example** Add: $\frac{1}{2} + \frac{3}{4} + \frac{15}{16} + \frac{11}{16}$

Express the fractions as equivalent fractions with 16 as the denominator.
Add the numerators.

Express $\frac{46}{16}$ as a mixed number in lowest terms.

$$\frac{1}{2} = \frac{1}{2} \times \frac{8}{8} = \frac{8}{16}$$
$$\frac{3}{4} = \frac{3}{4} \times \frac{4}{4} = \frac{12}{16}$$
$$\frac{15}{16} = \qquad\qquad \frac{15}{16}$$
$$+\frac{11}{16} = \qquad\qquad \frac{11}{16}$$
$$\rule{2cm}{0.4pt}$$
$$\frac{46}{16}$$

$$\frac{46}{16} = 2\frac{14}{16} = 2\frac{7}{8}$$

# MATH REVIEW 3
## Adding Combinations of Fractions, Mixed Numbers, and Whole Numbers

O To add mixed numbers or combinations of fractions, mixed numbers, and whole numbers, express the fractional parts of the numbers as equivalent fractions having the lowest common denominator. Add the whole numbers. Add the fractions. Combine the whole number and the fraction and express in lowest terms.

**Example 1** Add: $3\frac{7}{8} + 5\frac{1}{2} + 9\frac{3}{16}$

Express the fractional parts as equivalent fractions with 16 as the common denominator. Add the whole numbers. Add the fractions. Combine the whole number and the fraction. Express the answer in lowest terms.

$$3\frac{7}{8} = 3\frac{14}{16}$$
$$5\frac{1}{2} = 5\frac{8}{16}$$
$$+9\frac{3}{16} = 9\frac{3}{16}$$
$$\rule{3cm}{0.4pt}$$
$$17\frac{25}{16} = 17 + 1\frac{9}{16} = 18\frac{9}{16}$$

**Example 2** Add: $6\frac{3}{4} + \frac{9}{16} + 7\frac{21}{32} + 15$

Express the fractional parts as equivalent fractions with 32 as the common denominator. Add the whole numbers. Add the fractions. Combine the whole number and the fraction. Express the answer in lowest terms.

$$6\frac{3}{4} = 6\frac{24}{32}$$
$$\frac{9}{16} = \frac{18}{32}$$
$$7\frac{21}{32} = 7\frac{21}{32}$$
$$+15 = \qquad 15$$
$$\rule{3cm}{0.4pt}$$
$$28\frac{63}{32} = 28 + 1\frac{31}{32} = 29\frac{31}{32}$$

# MATH REVIEW 4
## Subtracting Fractions from Fractions

O Fractions must have a common denominator in order to be subtracted.

O To subtract a fraction from a fraction, express the fractions as equivalent fractions having the lowest common denominator. Subtract the numerators. Write their difference over the common denominator.

**Example** Subtract $\frac{3}{4}$ from $\frac{15}{16}$
Express the fractions as equivalent fractions with 16 as the common denominator. Subtract the numerator 12 from the numerator 15. Write the difference 3 over the common denominator 16.

$$
\begin{array}{r}
\frac{15}{16} = \frac{15}{16} \\
-\frac{3}{4} = -\frac{12}{16} \\
\hline
\frac{3}{16}
\end{array}
$$

# MATH REVIEW 5
## Subtracting Fractions and Mixed Numbers from Whole Numbers

O To subtract a fraction or a mixed number from a whole number, express the whole number as an equivalent mixed number. The fraction of the mixed number has the same denominator as the denominator of the fraction that is subtracted. Subtract the numerators of the fractions and write their difference over the common denominator. Subtract the whole numbers. Combine the whole number and fraction. Express the answer in lowest terms.

**Example 1** Subtract $\frac{3}{8}$ from 7.
Express the whole number as an equivalent mixed number with the same denominator as the denominator of the fraction that is subtracted ($7 = 6\frac{8}{8}$).
Subtract $\frac{3}{8}$ from $\frac{8}{8}$
Combine whole number and fraction.

$$
\begin{array}{r}
7 = 6\frac{8}{8} \\
-\frac{3}{8} = -\frac{3}{8} \\
\hline
6\frac{5}{8}
\end{array}
$$

**Example 2** Subtract $5\frac{15}{32}$ from 12
Express the whole number as an equivalent mixed number with the same denominator as the denominator of fraction that is subtracted ($12 = 11\frac{32}{32}$).
Subtract fractions.
Subtract whole numbers.
Combine whole number and fraction.

$$
\begin{array}{r}
12 = 11\frac{32}{32} \\
-5\frac{15}{32} = -5\frac{15}{32} \\
\hline
6\frac{17}{32}
\end{array}
$$

# MATH REVIEW 6
## Subtracting Fractions and Mixed Numbers from Mixed Numbers

O To subtract a fraction or a mixed number from a mixed number, the fractional part of each number must have the same denominator. Express fractions as equivalent fractions having a common denominator. When the fraction subtracted is larger than the fraction from which it is subtracted, one unit of a whole number is expressed as a fraction with the common denominator. Combine the whole number and fractions. Subtract fractions and subtract whole numbers.

**Example 1** Subtract $\frac{7}{8}$ from $4\frac{3}{16}$

Express the fractions as equivalent fractions with the common denominator 16. Since 14 is larger than 3, express one unit of $4\frac{3}{16}$ as a fraction and combine whole number and fractions.

$(4\frac{3}{16} = 3 + \frac{16}{16} + \frac{3}{16} = 3\frac{19}{16})$.

Subtract.

$$4\frac{3}{16} = 4\frac{3}{16} = 3\frac{19}{16}$$
$$-\frac{7}{8} = \frac{14}{16} = -\frac{14}{16}$$
$$3\frac{5}{16}$$

**Example 2** Subtract 13 from 20

Express the fractions as equivalent fractions with the common denominator 32.

Subtract fractions.

Subtract whole numbers.

$$20\frac{15}{32} = 20\frac{15}{32}$$
$$-13\frac{1}{4} = -13\frac{8}{32}$$
$$7\frac{7}{32}$$

# MATH REVIEW 7
## Multiplying Fractions

○ To multiply two or more fractions, multiply the numerators. Multiply the denominators. Write as a fraction with the product of the numerators over the product of the denominators. Express the answer in lowest terms.

**Example 1** Multiply $\frac{3}{4} \times \frac{5}{8}$

Multiply the numerators.

Multiply the denominators.

Write as a fraction.

$$\frac{3}{4} \times \frac{5}{8} = \frac{15}{32}$$

**Example 2** Multiply

Multiply the numerators.

Multiply the denominators.

Write as a fraction and express answer in lowest terms.

$$\frac{1}{2} \times \frac{2}{3} \times \frac{4}{5} = \frac{8}{30} = \frac{4}{15}$$

# MATH REVIEW 8
## Multiplying any Combination of Fractions, Mixed Numbers, and Whole Numbers

○ To multiply any combination of fractions, mixed numbers, and whole numbers, write the mixed numbers as fractions. Write whole numbers over the denominator 1. Multiply numerators. Multiply denominators. Express the answer in lowest terms.

──────────○

**Example 1** Multiply $3\frac{1}{4} \times \frac{3}{8}$

Write the mixed number $3\frac{1}{4}$ as the fraction $\frac{13}{4}$.
Multiply the numerators.
Multiply the denominators.
Express as a mixed number.

$$3\frac{1}{4} \times \frac{3}{8} = \frac{13}{4} \times \frac{3}{8} = \frac{39}{32} = 1\frac{7}{32}$$

──────────○

**Example 2** Multiply $2\frac{1}{3} \times 4 \times \frac{4}{5}$

Write the mixed number $2\frac{1}{3}$ as the fraction $\frac{7}{3}$.
Write the whole number 4 over 1.
Multiply the numerators.
Multiply the denominators.
Express as a mixed number.

$$2\frac{1}{3} \times 4 \times \frac{4}{5} = \frac{7}{3} \times \frac{4}{1} \times \frac{4}{5} = \frac{112}{15}$$

$$\frac{112}{15} = 7\frac{7}{15}$$

# MATH REVIEW 9
## Dividing Fractions

○ Division is the inverse of multiplication. Dividing by 4 is the same as multiplying by $\frac{1}{4}$. 4 is the inverse of $\frac{1}{4}$ and is the inverse of 4.
The inverse of $\frac{5}{16}$ is $\frac{16}{5}$.

○ To divide fractions, invert the divisor, change to the inverse operation and multiply. Express the answer in lowest terms.

──────────○

**Example** Divide: $\frac{7}{8} \div \frac{2}{3}$

Invert the divisor $\frac{2}{3}$
$\frac{2}{3}$ inverted is $\frac{3}{2}$.
Change to the inverse operation and multiply.
Express as a mixed number.

$$\frac{7}{8} \div \frac{2}{3} = \frac{7}{8} \times \frac{3}{2} = \frac{21}{16} = 1\frac{5}{16}$$

# MATH REVIEW 10
## Dividing any Combination of Fraction, Mixed Numbers, and Whole Numbers

○ To divide any combination of fractions, mixed numbers, and whole numbers, write the mixed number as fractions. Write whole numbers over the denominator 1. Invert the divisor. Change to the inverse operation and multiply. Express the answer in lowest terms.

**Example 1** Divide: $6 \div \frac{7}{10}$

Write the whole number 6 over the denominator 1.

Invert the divisor $\frac{7}{10}$; $\frac{7}{10}$ inverted is $\frac{10}{7}$.

Change to the inverse operation and multiply.

Express as a mixed number.

$$\frac{6}{1} \div \frac{7}{10} =$$

$$\frac{6}{1} \times \frac{10}{7} = \frac{60}{7} = 8\frac{4}{7}$$

**Example 2** Divide: $\frac{3}{4} \div 2\frac{1}{5}$

Write the mixed number divisor $2\frac{1}{5}$ as the fraction $\frac{11}{5}$.

Invert the divisor $\frac{11}{5}$; $\frac{11}{5}$ inverted is $\frac{5}{11}$.

Change to the inverse operation and multiply.

$$\frac{3}{4} \div \frac{11}{5} =$$

$$\frac{3}{4} \times \frac{5}{11} = \frac{15}{44}$$

**Example 3** Divide: $4\frac{5}{8} \div 7$

Write the mixed number $4\frac{5}{8}$ as the fraction $\frac{37}{8}$.

Write the whole number divisor over the denominator 1.

Invert the divisor $\frac{7}{1}$; $\frac{7}{1}$ inverted is $\frac{1}{7}$.

Change to the inverse operation and multiply.

$$\frac{37}{8} \div \frac{7}{1} =$$

$$\frac{37}{8} \times \frac{1}{7} = \frac{37}{56}$$

# MATH REVIEW 11
## Rounding Decimal Fractions

○ To round a decimal fraction, locate the digit in the number that gives the desired number of decimal places. Increase that digit by 1 if the digit that directly follows is 5 or more. Do not change the value of the digit if the digit that follows is less than 5. Drop all digits that follow.

**Example 1** Round 0.63861 to 3 decimal places.

Locate the digit in the third place (8). The fourth decimal-place digit, 6, is greater than 5 and increases the third decimal-place digit 8, to 9. Drop all digits that follow.

$$0.638\underline{6}1 \approx 0.639$$

**Example 2** Round 3.0746 to 2 decimal places.
Locate the digit in the second decimal place (7). The third decimal-place digit 4 is less than 5 and does not change the value of the second decimal-place digit 7. Drop all digits that follow.

$3.0\underline{7}46 \approx 3.07$

# MATH REVIEW 12
## Adding Decimal Fractions

○ To add decimal fractions, arrange the numbers so that the decimal points are directly under each other. The decimal point of a whole number is directly to the right of the last digit. Add each column as with whole numbers. Place the decimal point in the sum directly under the other decimal points.

**Example** Add: $7.65 + 208.062 + 0.009 + 36 + 5.1037$
Arrange the numbers so that the decimal points are directly under each other.
Add zeros so that all numbers have the same number of places to the right of the decimal point.
Add each column of numbers.
Place the decimal point in the sum directly under the other decimal points.

```
    7.6500
  208.0620
    0.0090
   36.0000
+   5.1037
  256.8247
```

# MATH REVIEW 13
## Subtracting Decimal Fractions

○ To subtract decimal fractions, arrange the numbers so that the decimal points are directly under each other. Subtract each column as with whole numbers. Place the decimal point in the difference directly under the other decimal points.

**Example** Subtract: $87.4 - 42.125$
Arrange the numbers so that the decimal points are directly under each other. Add zeros so that the numbers have the same number of places to the right of the decimal point.
Subtract each column of numbers.
Place the decimal point in the difference directly under the other decimal points.

```
  87.400
- 42.125
  45.275
```

# MATH REVIEW 14
## Multiplying Decimal Fractions

○ To multiply decimal fractions, multiply using the same procedure as with whole numbers. Count the number of decimal places in both the multiplier and the multiplicand. Begin counting from the last digit on the right of the product and place the decimal point the same number of places as there are in both the multiplicand and the multiplier.

──────────○

**Example** Multiply: $50.216 \times 1.73$
Multiply as with whole numbers.
Count the number of decimal places in the multiplier (2 places) and the multiplicand (3 places).
Beginning at the right of the product, place the decimal point the same number of places as there are in both the multiplicand and the multiplier (5 places).

$$
\begin{array}{r}
50.216 \quad \text{Multiplicand (3 places)} \\
\times\ 1.73 \quad \text{Multiplier (2 places)} \\
\hline
150648 \\
351512 \\
50216 \\
\hline
86.87368 \quad \text{(5 places)}
\end{array}
$$

──────●──────

○ When multiplying certain decimal fractions, the product has a smaller number of digits than the number of decimal places required. For these products, add as many zeros to the left of the product as are necessary to give the required number of decimal places.

──────────○

**Example** Multiply: $0.27 \times 0.18$
Multiply as with whole numbers.
The product must have 4 decimal places.
Add one zero to the left of the product.

$$
\begin{array}{r}
0.27 \ \text{(2 places)} \\
\times 0.18 \ \text{(2 places)} \\
\hline
216 \\
27 \\
\hline
0.0486 \ \text{(4 places)}
\end{array}
$$

──────●──────

# MATH REVIEW 15
## Dividing Decimal Fractions

○ To divide decimal fractions, use the same procedure as with whole numbers. Move the decimal point of the divisor as many places to the right as necessary to make the divisor a whole number. Move the decimal point of the dividend the same number of places to the right. Add zeros to the dividend if necessary. Place the decimal point in the answer directly above the decimal point in the dividend. Divide as with whole numbers. Zeros may be added to the dividend to find the number of decimal places required in the answer.

**Example 1** Divide: 0.6150 ÷ 0.75

Move the decimal point 2 places to the right in the divisor.
Move the decimal point 2 places in the dividend.
Place the decimal point in the answer directly above the decimal point in the dividend.
Divide as with whole numbers.

$$\text{Divisor} \longrightarrow 0\,75.\,\overline{)\,0\,61.50}\longleftarrow \text{Dividend}$$

```
            0.82
0 75. )0 61.50
       60 0
        1 50
        1 50
```

**Example 2** Divide: 10.7 ÷ 4.375. Round the answer to 3 decimal places.

Move the decimal point 3 places to the right in the divisor.
Move the decimal point 3 places in the dividend, adding 2 zeros.
Place the decimal point in the answer directly above the decimal point in the dividend.
Add 4 zeros to the dividend. One more zero is added than the number of decimal places required in the answer.
Divide as with whole numbers.

```
              2.4457 ≈ 2.446
4 375. )10 700.0000
        8 750
        1 950 0
        1 750 0
          200 00
          175 00
           25 00
           21 875
            3 1250
            3 0625
              625
```

# MATH REVIEW 16
## Expressing Comon Fractions as Decimal Fractions

○ A common fraction is an indicated division. A common fraction is expressed as a decimal fraction by dividing the numerator by the denominator.

**Example** Express as a decimal fraction.

Write as an indicated division.

Place a decimal point after the 5 and add zeros to the right of the decimal point.

Place the decimal point for the answer directly above the decimal point in the dividend.

Divide.

```
8)5
8)5.000
8)5.000

  0.625
8)5.000
```

○ A common fraction which will not divide evenly is expressed as a repeating decimal.

**Example** Express $\frac{1}{3}$ as a decimal fraction.

Write $\frac{1}{3}$ as an indicated division.

Place a decimal point after the 1 and add zeros to the right of the decimal point.

Place the decimal point for the answer directly above the decimal point in the dividend.

Divide.

```
3)1
3)1.0000
3)1.0000

  0.3333
3)1.0000
```

# MATH REVIEW 17
## Expressing Decimal Fractions as Common Fractions

○ To express a decimal fraction as a common fraction, write the number after the decimal point as the numerator of a common fraction. Write the denominator as 1 followed by as many zeros as there are digits to the right of the decimal point. Express the common fraction in lowest terms.

**Example 1** Express 0.9 as a common fraction.
Write 9 as the numerator.
Write the denominator as 1 followed by 1 zero. The denominator is 10.

$$\frac{9}{10}$$

**Example 2** Express 0.125 as a common fraction.
Write 125 as the numerator.
Write the denominator as 1 followed by 3 zeros. The denominator is 1000.
Express the fraction in lowest terms.

$$\frac{125}{1000}$$

$$\frac{125}{1000} = \frac{1}{8}$$

# MATH REVIEW 18
## Expressing Inches as Feet and Inches

○ There are 12 inches in 1 foot.

○ To express inches as feet and inches, divide the given length in inches by 12 to obtain the number of whole feet. The remainder is the number of inches in addition to the number of whole feet. The answer is the number of whole feet plus the remainder in inches.

```
        14   (feet)
12)176 7/16
    12
    56
    48
     8 7/16  ← Remainder
              (feet)
```

**Example 1** Express $176\frac{7}{16}$ inches as feet and inches.
Divide $176\frac{7}{16}$ inches by 12.
There are 14 feet plus a remainder of $8\frac{7}{16}$ inches.

$$14' - 8\frac{7}{16}\,''$$

**Example 2** Express 54.2 inches as feet and inches.
Divide 54.2 inches by 12.
There are 4 feet plus a remainder of 6.2 inches.

```
         4   (feet)
12) 54.2
    48
    6.2  ← Remainder
          (inches)
```

4 feet 6.2 inches

# MATH REVIEW 19
## Expressing Feet and Inches as Inches

○ There are 12 inches in 1 foot.

○ To express feet and inches as inches, multiply the number of feet in the given length by 12. To this product, add the number of inches in the given length.

───────────○

**Example** Express 7 feet $9\frac{3}{4}$ inches as inches.
Multiply 7 feet by 12. There are 84 inches in 7 feet.
Add $9\frac{3}{4}$ inches to 84 inches.

$7 \times 12 = 84$
7 feet = 84 inches
84 inches + $9\frac{3}{4}$ inches =
$93\frac{3}{4}$ inches

●─────────────

# Math Review 20
## Expressing Inches as Decimal Fractions of a Foot

○ An inch is $\frac{1}{12}$ of a foot. To express whole inches as a decimal part of a foot, divide the number of inches by 12.

───────────○

**Example** Express 7 inches as a decimal fraction of a foot.
Divide 7 by 12.

$7 \div 12 = 0.58$
0.58 feet

●─────────────

○ To express a common fraction of an inch as a decimal fraction of a foot, express the common fraction as a decimal, then divide the decimal by 12.

───────────○

**Example 1** Express $\frac{3}{4}$ inch as a decimal fraction of a foot.
Express $\frac{3}{4}$ as a decimal.
Divide the decimal by 12.

$3 \div 4 = 0.75$
$0.75 \div 12 = 0.06$
0.06 feet

●─────────────

───────────○

**Example 2** Express $4\frac{3}{4}$ inches as a decimal fraction of a foot.
Express $4\frac{3}{4}$ as a decimal.
Divide the decimal inches by 12.

$4 + \frac{3}{4} = 4 + 0.75 = 4.75$
$4.75 \div 12 = 0.39$
0.39 feet

●─────────────

# MATH REVIEW 21
## Expressing Decimal Fractions of a Foot as Inches

○ To express a decimal part of a foot as decimal inches, multiply by 12.

**Example:** Express 0.62 foot as inches.
Multiply 0.62 by 12.

$0.62 \times 12 = 7.44$
7.44 inches

○ To express a decimal fraction of an inch as a common fraction, see Math
Review 17.

# MATH REVIEW 22
## Area Measure

○ A surface is measured by determining the number of surface units contained in it. A surface is two-dimensional. It has length and width but no thickness. Both length and width must be expressed in the same unit of measure. Area is expressed in square units. For example, 5 feet 3 8 feet equals 40 square feet.

○ *Equivalent Units of Area Measure:*
  1 square foot (sq ft) = 12 inches $\times$ 12 inches = 144 square inches (sq in)
  1 square yard (sq yd) = 3 feet $\times$ 3 feet = 9 square feet (sq ft)

○ To express a given unit of area as a larger unit of area, divide the given area by the number of square units contained in one of the larger units.

**Example 1** Express 648 square inches as square feet.
Since 144 sq in = 1 sq ft, divide 648 by 144.

$648 \div 144 = 4.5$
648 square inches = 4.5 square feet

**Example 2** Express 28.8 square feet as square yards.
Since 9 sq ft = 1 sq yd, divide 28.8 by 9.

$28.8 \div 9 = 3.2$
28.8 square feet = 3.2 square yards

○ To express a given unit of area as a smaller unit of area, multiply the given area by the number of square units contained in one of the larger units.

**Example 1** Express 7.5 square feet as square inches.
Since 144 sq in = 1 sq ft, divide 7.5 by 144.

$7.5 \times 144 = 1080$
23 square yards = 207 square feet

**Example 2** Express 23 square yards as square feet.
Since 9 sq ft = 1 sq yd, multiply 23 by 9.

$23 \times 9 = 207$
23 square yards = 207 square feet

○ *Computing Areas of Common Geometric Figures:*

  1. **Rectangle** A rectangle is a four-sided plane figure with four right (90°) angles.

The area of a rectangle is equal to the product of its length and its width.
Area = length × width (A = l × w)

LENGTH (l)

_____○

**Example** Find the area of a rectangle 24 feet long and 13 feet wide.
A = l × w
A = 24 ft × 13 ft
A = 312 square feet

●_____

2. **Triangle** A triangle is a plane figure with three sides and three-angles. The area of a triangle is equal to one-half the product of its base and altitude.

$A = \frac{1}{2}$ base × altitude (A = b × a)

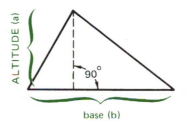
base (b)

_____○

**Example** Find the area of a triangle with a base of 16 feet and an altitude of 12 feet.
$A = \frac{1}{2}$ b × a
$A = \frac{1}{2}$ × 16 ft × 12 ft
A = 96 square feet

●_____

3. **Circle** The area of a circle is equal to p times the square of its radius.
Area = π × radius² (A = π × r²)
Note: π (pronounced "pi") is approximately equal to 3.14.
Radius squared (r²) means r × r.

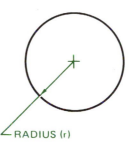
RADIUS (r)

_____○

**Example** Find the area of a circle with a 15-inch radius.
A = π × r²
A = 3.14 × (15 in)²
A = 3.14 × 225 sq in
A = 706.5 square inches

●_____

# MATH REVIEW 23
## Volume Measure

○ A solid is measured by determining the number of cubic units contained in it. A solid is three-dimensional; it has length, width, and thickness or height. Length, width, and thickness must be expressed in the same unit of measure. Volume is expressed in cubic units. For example, 3 feet × 5 feet × 10 feet = 150 cubic feet.

○ *Equivalent Units of Volume Measure:*
1 cubic foot (cu ft) =
    12 in × 12 in × 12 in = 1728 cubic inches (cu in)
1 cubic yard (cu yd) =
    3 ft × 3 ft × 3 ft = 27 cubic feet (cu ft)

○ To express a given unit of volume as a larger unit of volume, divide the given volume by the number of cubic units contained in one of the larger units.

**Example 1** Express 6048 cubic inches as cubic feet.
Since 1728 cu in = 1 cu ft, divide 6048 by 1728.

$6048 \div 1728 = 3.5$
6048 cubic inches = 3.5 cubic feet

**Example 2** Express 167.4 cubic feet as cubic yards.
Since 17 cu ft = 1 cu yd, divide 167.4 by 27.

$167.4 \div 27 = 6.2$
167.4 cubic feet = 6.2 yards

○ To express a given unit of volume as a smaller unit of volume, multiply the given volume by the number of cubic units contained in one of the larger units.

**Example 1** Express 1.6 cubic feet as cubic inches.
Since 1728 cu in = 1 cu ft, divide 1.6 by 1728.

$1.6 \times 1728 = 2764.8$
1.6 cubic feet = 2764.8 cubic inches

**Example 2** Express 8.1 cubic yards as cubic feet.
Since 27 cu ft = 1 cu yd, divide 8.1 by 27.

$8.1 \times 27 = 218.7$
8.1 cubic yards = 218.7 cubic feet

○ *Computing Volumes of Common Solids*

○ A prism is a solid that has two identical faces called bases and parallel lateral edges. In a right prism, the lateral edges are perpendicular (at 90°) to the bases. The altitude or height (h) of a prism is the perpendicular distance between its two bases. Prisms are named according to the shapes of their bases.

○ The volume of any prism is equal to the product of the area of its base and altitude or height.
Volume = area of base 3 altitude (V = AB × h)

○ *Right Rectangular Prism:*
A right rectangular prism has rectangular bases.
Volume = area of base 3 altitude
    $V = A_B \times h$

**Example** Find the volume of a rectangular prism with a base length of 20 feet, a base width of 14 feet, and a height (altitude) of 8 feet.

$V = A_B \times h$

Compute the area of the base $(A_B)$:

Area of base = length $\times$ width

$A_B$ = 20 ft $\times$ 14 ft

$A_B$ = 280 sq ft

Compute the volume of the prism:

$V = A_B \times h$

$V$ = 280 sq ft $\times$ 8 ft

$V$ = 2240 cu ft

○ *Right Triangular Prism:*

A right triangular prism has triangular bases.

Volume = area of base $\times$ altitude

$V = A_B \times h$

**Example** Find the volume of a triangular prism in which the base of the triangle is 5 feet, the altitude of the triangle is 3 feet, and the altitude (height) of the prism is 4 feet. Refer to the accompanying figure.

Volume = area of base $\times$ altitude

$V = A_B \times h$

Compute the area of the base:

Area of base = $\frac{1}{2}$ base of triangle $\times$ altitude of triangle

$A_B = \frac{1}{2} b \times a$

$A_B = \frac{1}{2} \times 5$ ft $\times$ 3 ft

$A_B$ = 7.5 sq ft

Compute the volume of the prism:

$V = A_B \times h$

$V$ = 7.5 sq ft $\times$ 4 ft

$V$ = 30 cubic feet

○ *Right Circular Cylinder:*

A right circular cylinder has circular bases.

Volume = area of base $\times$ altitude

$V = A_B \times h$

**Example** Find the volume of a circular cylinder 1 foot in diameter and 10 feet high.

Note: Radius = $\frac{1}{2}$ Diameter; Radius = $\frac{1}{2} \times 1$ ft = 0.5 ft.

$V = A_B \times h$

Compute the area of the base:

Area of base = $\pi \times$ radius squared

$A_B = 3.14 \times (0.5)^2$

$A_B = 3.14 \times 0.5$ ft $\times$ 0.5 ft

$A_B = 3.14 \times 0.25$ sq ft

$A_B = 0.785$ sq ft

Compute the volume of the cylinder:

$V = A_B \times h$

$V = 0.785 \text{ sq ft} \times 10 \text{ ft}$

$V = 7.85 \text{ cubic feet}$

# MATH REVIEW 24
## Finding An Unknown Side of a Right Triangle, Given Two Sides

○ If one of the angles of a triangle is a right (90°) angle, the figure is called a right triangle. The side opposite the right angle is called the hypotenuse. In the figure shown, c is opposite the right angle; c is the hypotenuse.

○ In a right triangle, the square of the hypotenuse is equal to the sum of the squares of the other two sides:

$C^2 = a^2 + b^2$

If any two sides of a right triangle are known, the length of the third side can be determined by one of the following formulas:

$c = \sqrt{a^2 + b^2}$

$a = \sqrt{c^2 - b^2}$

$b = \sqrt{c^2 - a^2}$

**Example 1**  In the right triangle shown, a = 6 ft, b = 8 ft, find c.

$c = \sqrt{a^2 + b^2}$

$c = \sqrt{6^2 + 8^2}$

$c = \sqrt{36 + 64}$

$c = \sqrt{100}$

$c = 10 \text{ feet}$

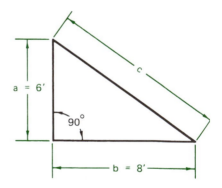

**Example 2**  In the right triangle shown, c = 30 ft, b = 20 ft, find a.

$a = \sqrt{c^2 - b^2}$

$a = \sqrt{30^2 - 20^2}$

$a = \sqrt{900 - 400}$

$a = \sqrt{500}$

$a = 22.36 \text{ feet (to 2 decimal places)}$

**Example 3** In the right triangle shown, c = 18 ft, a = 6 ft, find b.

$$b = \sqrt{c^2 - a^2}$$

$$b = \sqrt{18^2 - 6^2}$$

$$b = \sqrt{324 - 36}$$

$$b = \sqrt{288}$$

b = 16.97 feet (to 2 decimal places)

# MATERIAL SYMBOLS IN SECTIONS

EARTH

ROCK

GRAVEL OR
CRUSHED STONE

CONCRETE

CONCRETE BLOCK

FACE BRICK OR
COMMON BRICK

FIRE BRICK

ROUGH WOOD-CONTINUOUS

WOOD BLOCKING

FINISH WOOD

STRUCTURAL STEEL

REINFORCING BARS

GENERAL METAL

BATT
INSULATION

RIGID
INSULATION

PLASTER OR
GYPSUM BOARD

THIN SHEET MATERIALS
(PLASTIC FILM, SHEET
METAL, PAPER, ETC.)

# PLUMBING SYMBOLS

**PIPING**

DRAIN OR WASTE
ABOVE GROUND ────────────────────────

DRAIN OR WASTE
BELOW GROUND ── ── ── ── ── ──

VENT ‐ ‐ ‐ ‐ ‐ ‐ ‐ ‐ ‐ ‐ ‐ ‐

COLD WATER ──── ── ──── ── ────

HOT WATER ──── ‐ ── ‐ ── ‐ ──

HOT WATER
HEAT SUPPLY ───── HW ───── HW ─────

HOT WATER
HEAT RETURN ───── HWR ───── HWR ─────

GAS ───── G ───── G ─────

PIPE TURNING
DOWN OR AWAY ─────────────○

PIPE TURNING
UP OR TOWARD ─────────────○

BREAK—PIPE CONTINUES ─────────────/

| FITTINGS | SOLDERED | SCREWED |
|---|---|---|
| TEE | | |
| WYE | | |
| ELBOW – 90° | | |
| ELBOW – 45° | | |
| CAP | | |
| UNION CLEANOUT | | |
| STOP VALVE | | |

# ELECTRICAL SYMBOLS

| | |
|---|---|
| CEILING FIXTURE | |
| CEILING FIXTURE WITH PULL SWITCH | PS |
| WALL MOUNTED FIXTURE | |
| RECESSED CEILING FIXTURE – OUTLINE SHOWS SHAPE | |
| FLOURESCENT FIXTURE | |
| FAN OUTLET | F or |
| CONVENIENCE DUPLEX OUTLET | |
| SPLIT WIRED DUPLEX OUTLET | |
| WEATHERPROOF OUTLET | WP |
| OUTLET WITH GROUND FAULT INTERRUPTER | GFI |
| SPECIAL-EQUIPMENT OUTLET | |
| RANGE OUTLET | R |
| PUSH BUTTON | |
| CHIME | CH |
| TRANSFORMER | T |

TELEPHONE

INTERCOM

TELEVISION ANTENNA    TV

SMOKE DETECTOR

DISTRIBUTION PANEL

JUNCTION BOX    J

SINGLE-POLE SWITCH    S

THREE-WAY SWITCH    $S_3$

SWITCH WITH PILOT LIGHT    $S_p$

WEATHERPROOF SWITCH    $S_{wp}$

SWITCH WIRING    or

# ABBREVIATIONS

A.B.—anchor bolt
A.C.—air conditioning
AL. or ALUM.—aluminum
BA—bathroom
BLDG.—building
BLK.—block
BLKG.—blocking
BM.—beam
BOTT.—bottom
B.PL.—base plate
BR—bedroom
BRM.—broom closet
BSMT.—basement
CAB.—cabinet
℄—centerline
CLNG. or CLG.—ceiling
C.M.U.—concrete masonry unit
  (concrete block)
CNTR.—center or counter
COL.—column
COMP.—composition
CONC.—concrete
CONST.—construction
CONT.—continuous
CORRUG.—corrugated
CRNRS.—corners
CU—copper
d—penny (nail size)
DBL.—double
DET.—detail
DIA or ⌀—diameter
DIM.—dimension
DN.—down
DO—ditto
DP.—deep or depth
DR.—door
D.W.—dishwasher
ELEC.—electric
ELEV.—elevation
EQ.—equal
EXP.—exposed or expansion

EXT.—exterior
F.G.—fuel gas
FIN.—finish
FL. or FLR.—floor
FOUND. or FDN.—foundation
F.P.—fireplace
FT.—foot or feet
FTG.—footing
GAR.—garage
G.F.C.I.—ground fault circuit
  interrupter
G.I.—galvanized iron
GL.—glass
GRD.—grade
GYP.BD.—gypsum board
H.C.—hollow core door
H.C.W.—hollow core wood
HDR.—header
H.M.—hollow metal
HORIZ.—horizontal
HT. or HGT.—height
H.W.—hot water
H.W.M.—high water mark
IN.—inch or inches
INSUL.—insulation
INT.—interior
JSTS.—joists
JT.—joint
LAV.—lavatory
L.H.—left hand
LIN.—linen closet
LT.—light
MANUF.—manufacturer
MAS.—masonry
MATL.—material
MAX.—maximum
MIN.—minimum
MTL.—metal
NAT.—natural
N/F—now or formerly
N.I.C.—not in contract

o/—overhead or over
O.C.—on centers
O.H. Door—overhead door
O.S.B.—Oriented strand
  board
PERF.—perforated
PL—plate
PLYWD.—plywood
P.T.—pressure-treated lumber
R—risers
REF.—refrigerator
REINF.—reinforcement
REQ.—requirement
R.H.—right hand
RM—room
R.O.B.—run of bank (gravel)
R.O.W.—right of way
SCRND.—screened
SHT.—sheet
SHTG.—sheathing
SHWR.—shower
SIM.—similar
SL.—sliding
S&P—shelf and pole
SQ. or �少 —square
STD.—standard
STL.—steel
STY.—story
T&G—tongue and groove
THK.—thick
T'HOLD.—threshold
TYP.—typical
V.B.—vapor barrier
w/—with
WARD.—wardrobe
W.C.—water closet
WD.—wood
WDW.—window
W.H.—water heater
W.I.—wrought iron

# Index